SpringerBriefs in Molecular Science

Ultrasound and Sonochemistry

W0080381

Series editors

Bruno G. Pollet, Faculty of Engineering, Norwegian University of Science and Technology, Trondheim, Norway
Muthupandian Ashokkumar, School of Chemistry, University of Melbourne, Melbourne, VIC, Australia

SpringerBriefs in Molecular Science: Ultrasound and Sonochemistry is a series of concise briefs that present those interested in this broad and multidisciplinary field with the most recent advances in a broad array of topics. Each volume compiles information that has thus far been scattered in many different sources into a single, concise title, making each edition a useful reference for industry professionals, researchers, and graduate students, especially those starting in a new topic of research.

More information about this series at http://www.springer.com/series/15634

About the Series Editors

Bruno G. Pollet is a full Professor of Renewable Energy at the Norwegian University of Science and Technology (NTNU) in Trondheim. He is a Fellow of the *Royal Society of Chemistry* (RSC), an Executive Editor of *Ultrasonics Sonochemistry* and a Board of Directors' member of the *International Association of Hydrogen Energy* (IAHE). He held Visiting Professorships at the University of Ulster, Professor Molkov's HySAFER (UK) and at the University of Yamanashi, Professor Watanabe's labs (Japan). His research covers a wide range of areas in Electrochemical Engineering, Electrochemical Energy Conversion and Sono-electrochemistry (Power Ultrasound in Electrochemistry) from the development of novel materials, hydrogen and fuel cell to water treatment/disinfection demonstrators & prototypes. He was a full Professor of Energy Materials and Systems at the University of the Western Cape (South Africa) and R&D Director of the National Hydrogen South Africa (HySA) Systems Competence Centre. He was also a Research Fellow and Lecturer in Chemical Engineering at The University of Birmingham (UK) as well as a co-founder and an Associate Director of The University of Birmingham Centre for Hydrogen and Fuel Cell Research. He has worked for Johnson Matthey Fuel Cells Ltd (UK) and other various industries worldwide as Technical Account Manager, Project Manager, Research Manager, R&D Director, Head of R&D and Chief Technology Officer. He was awarded a Diploma in Chemistry and Material Sciences from the Université Joseph Fourier (Grenoble, France), a B.Sc. (Hons) in Applied Chemistry from Coventry University (UK) and an M.Sc. in Analytical Chemistry from The University of Aberdeen (UK). He also gained his Ph.D. in Physical Chemistry in the field of Electrochemistry and Sonochemistry under the supervision of Profs. J. Phil Lorimer & Tim J. Mason at the Sonochemistry Centre of Excellence, Coventry University (UK). He undertook his PostDoc in Electrocatalysis at the Liverpool University Electrochemistry group led by Prof. David J. Schiffrin. Bruno has published many scientific publications, articles, book chapters and books in the field of Sonoelectrochemistry, Fuel Cells, Electrocatalysis and Electrochemical Engineering. Bruno is member of editorial board journals (*International Journal of Hydrogen Energy*/*Electrocatalysis*/*Ultrasonics Sonochemistry*/*Renewables-Wind, Water and Solar*/*Electrochem*). He is also fluent in English, French and Spanish. *Current Editorships: Hydrogen Energy and Fuel Cells Primers Series (AP, Elsevier) and Ultrasound and Sonochemistry (Springer)*.

Prof. Muthupandian Ashokkumar (Ashok) is a Physical Chemist who specializes in Sonochemistry, teaches undergraduate and postgraduate Chemistry and is a senior academic staff member of the School of Chemistry, University of Melbourne. Ashok is a renowned sonochemist, with more than 20 years of experience in this field, and has developed a number of novel techniques to characterize acoustic cavitation bubbles and has made major contributions of applied sonochemistry to the Materials, Food and Dairy industry. His research team has developed a novel ultrasonic processing technology for improving the functional properties of dairy ingredients. Recent research also involves the ultrasonic synthesis of functional nano- and biomaterials that can be used in energy production, environmental remediation and diagnostic and therapeutic medicine. He is the Deputy Director of an Australian Research Council Funded Industry Transformation Research Hub (ITRH; http://foodvaluechain.unimelb.edu.au/#research; Industry Partner: Mondelez International) and leading the Encapsulation project (http://foodvaluechain.unimelb.edu.au/research/ultrasonic-encapsulation). He has received about $ 15 million research grants to support his research work that includes several industry projects. He is the Editor-in-Chief of *Ultrasonics Sonochemistry*, an international journal devoted to sonochemistry research with a Journal Impact Factor of 4.3). He has edited/co-edited several books and special issues for journals; published ~360 refereed papers (H-Index: 49) in high impact international journals and books; and delivered over 150 invited/keynote/plenary lectures at international conferences and academic institutions. Ashok has successfully organised 10 national/international scientific conferences/workshops and managed a number of national and international competitive research grants. He has served on a number of University of Melbourne management committees and scientific advisory boards of external scientific organizations. Ashok is the recipient of several prizes, awards and fellowships, including the Grimwade Prize in Industrial Chemistry. He is a Fellow of the RACI since 2007.

Jayani Chandrapala · Bogdan Zisu

Ultrasound Technology in Dairy Processing

 Springer

Jayani Chandrapala
Department of Biosciences and Food
Technology, School of Science
RMIT University
Melbourne, VIC, Australia

Bogdan Zisu
Department of Biosciences and Food
Technology, School of Science
RMIT University
Melbourne, VIC, Australia

ISSN 2191-5407   ISSN 2191-5415   (electronic)
SpringerBriefs in Molecular Science
ISSN 2511-123X   ISSN 2511-1248   (electronic)
Ultrasound and Sonochemistry
ISBN 978-3-319-93481-5   ISBN 978-3-319-93482-2   (eBook)
https://doi.org/10.1007/978-3-319-93482-2

Library of Congress Control Number: 2018945455

Printed on acid-free paper

This Springer imprint is published by the registered company Springer International Publishing AG part of Springer Nature
The registered company address is: Gewerbestrasse 11, 6330 Cham, Switzerland

Contents

Chapter 1
Ultrasound Technology in Dairy Processing

Abstract High-intensity ultrasound technology has been vastly utilized as a processing method in a number of dairy applications in preference to traditional thermal treatments in recent years. Acoustic cavitation generates physical forces such as acoustic streaming, acoustic radiation, shear, micro-jetting and shockwaves. These forces are utilized in specific dairy applications including emulsification, filtration, functionality modifications, microbial inactivation, homogenization, crystallization and the separation of fat. Although some of these applications are adopted by industry for large-scale operations, most are still limited to laboratory scale. Due to its widespread potential, it is becoming increasingly clear that ultrasound technology has huge potential as an energy efficient emerging technology across the dairy sector.

Keywords High-intensity ultrasound · Low-intensity ultrasound · Dairy Emulsification · Heat stability

1.1 Introduction—High-Intensity Ultrasound

Ultrasound (US) is identified as sound wave beyond the range of human hearing. Ultrasound has a frequency range of 20 kHz–10 MHz and is further divided into three main areas; low-frequency high-power ultrasound with intensities >1 W/cm^2, intermediate frequency medium power ultrasound and high-frequency low-power ultrasound with intensities <1 W/cm^2 (Ashokkumar and Mason 2007). The interaction between the ultrasonic waves, liquid and dissolved gas leads to 'acoustic cavitation' when ultrasound travels through a liquid medium. Dissolved gas bubbles will grow due to rectified diffusion and bubble–bubble coalescence. A dissolved gas bubble oscillates under the impact of ultrasound waves and experiences a fluctuating pressure. Dissolved gas and solvent vapor diffuse in and out of the oscillating bubbles depending on whether it is the expansion or the compression phase. There will be a net growth of the bubbles if the extent of gas/solvent vapor that diffuses into the bubbles during the expansion stage is more than the extent of gas/solvent vapor that diffuses out of the bubble during the compression stage. However, when

© The Author(s), under exclusive licence to Springer International Publishing AG, part of Springer Nature 2018
J. Chandrapala and B. Zisu, *Ultrasound Technology in Dairy Processing*, Ultrasound and Sonochemistry, https://doi.org/10.1007/978-3-319-93482-2_1

bubbles reach the resonance size range which depends on the ultrasonic frequency, they will grow to a maximum size and violently collapse. This collapse generates very high temperatures up to 10,000 K and pressure conditions. This is known as 'transient cavitation'. Furthermore, stable cavitation can occur where the resonance sized bubbles continue to oscillate for some time without collapsing (Ashokkumar 2011). Highly reactive radicals are generated along with intense physical forces such as shock waves, micro-jets, shear forces and turbulence. Transient cavitation events are less frequent at the intermediate frequency range of ~200–500 kHz. In contrast, strong physical effects result mainly from the transient cavitation at the lower frequency range of ~20–100 kHz.

1.2 Use of Ultrasound in Dairy Applications

1.2.1 Emulsification

The kinetic stability of thermodynamically unstable dairy emulsions can be attained with the addition of emulsifiers and through an efficient emulsifying method. In this context, ultrasound technology serves as an excellent candidate due to such advantages as production of smaller droplet sizes, narrow size distribution, ability to use minimal emulsifier contents, increased emulsion stability, easy operation and cleaning and low production costs due to low-energy requirements. Thus, low-frequency (<100 kHz) high-power (>10 Wcm^{-2}) ultrasound has been widely exploited for producing stable emulsions.

Shock waves created by a bubble collapse near the phase boundary of two immiscible liquids lead to efficient mixing of the two liquids. US emulsification is described as a two-step process (Canselier et al. 2002). The first step involves the explosion of the dispersed phase droplets into the continuous phase leading to the formation of smaller droplets with the use of the turbulence caused by mechanical vibration. The second step is the breakup of the droplets through the shear forces generated by cavitation at the interface initiating droplet–droplet coalescence. The final droplet-size distribution obtained during emulsification results due to the competition between these two opposite processes (Vankova et al. 2007). The US energy not only breaks the planar interface but also overcomes the Laplace pressure in order to produce finer droplets where large amounts of shear are required to overcome the interfacial tension in between liquids (Canselier et al. 2002). However, Shanmugam and Ashokkumar (2014a) stated that milk components lower the interfacial tension which in turn lowers the Laplace pressure, and hence lower amount of shear is required to break the droplets.

Emulsification by ultrasound is influenced by applied power (Abismail et al. 2000; Freitas et al. 2006), position of the ultrasonic source with liquid–liquid interface (Sivakumar et al. 2002), tip diameter (Juang and Lin 2004), vessel size (Juang and Lin 2004), viscosity and the composition of the continuous phase (O'Sullivan et al.

2014), surfactants concentration (Abismail et al. 1999), hydrostatic pressure and presence of dissolved gases (Behrund and Schubert 2001).

Shanmugam and Ashokkumar (2014a) demonstrated that incorporation of 7% flaxseed oil into pasteurized homogenized skim milk (PHSM) using high-intensity US operated at 20 kHz for 1–8 min with varying power levels (88, 132 and 176 W) stabilized emulsions. Increasing power and residence time decreased the emulsion droplet size. An increase in power also increased the size and the number of cavitation bubbles (Ashokkumar and Mason 2007) which in turn increased the intensity of the bubble collapse resulting in high shear forces leading to smaller droplets. Sonication at an applied power of 176 W for 3 min was sufficient to produce flaxseed oil: milk emulsion droplets with an average mean volume diameter of 0.64 μm with increased stability for up to 9 days. A small proportion (<20%) of partially denatured whey proteins provided the stability to the emulsion droplets with no addition of surfactants. Further, minimum process times of 3 and 6 min were recommended in manufacturing 15 and 21% high oil emulsions at 176 W. In contrast to US, mechanical mixing with an Ultraturrax (UT) at similar energy densities did not produce stable emulsions even after 20 min of processing highlighting the differences between the technologies. Thus, US was capable of inducing structural changes within the protein molecules allowing them to more effectively stabilize the interfaces (Shanmugam et al. 2012). In some instances, US assists protein cross-linking and facilitate aggregation (Cavalieri et al. 2008).

Similarly, Nejadmansouri et al. (2016) created WPI-stabilized fish oil nano-emulsions with small droplets and with thick protein layers at the interface after sonication at 200 W for 20-min. However, no further reductions were observed after 20 min indicative of an energy threshold limit for US size reduction of droplets (Calligaris et al. 2016). In addition, they found a decrease oxidation rate for WPI-stabilized nano-emulsions and an increase in the antioxidant capacities of the WPI fractions.

O'Sullivan et al. (2014) studied the emulsifying properties of sodium caseinate, WPI and MPI sonicated at 34 Wcm^{-2} for 2 min. The emulsions made with sodium caseinate and WPI did not show significant size reductions with treatment at all concentrations, while MPI showed smaller droplet sizes at concentrations below 1 wt%. This was attributed to the significant reduction of the casein micelle size and increase in hydrophobicity in MPI. This, in turn, increased the rate of protein adsorption to the oil-in-water interface and reduced the interfacial tension which facilitated the droplet breakages. However, the US-mediated emulsions exhibited considerable stability over time. Another study by O'Sullivan et al. (2015a, b) investigated the effect of processing volume (3–13 mL), residence time (1–300 s), amplitude (20–40%) and the concentration of the emulsifier (0.1–3 wt%) at lab scale (120–453 Wcm^{-2}) and pilot scale (20–62 Wcm^{-2}) using MPI as the emulsifier. It was found that an increase in residence time, decrease in volume and increase in amplitude decreased the emulsion droplet size and improved the emulsification efficiency. Similarly, Furtado et al. (2017) showed reduced emulsion droplet size, narrow size distribution and increased stability for sodium caseinate emulsions sonicated at 300 W for 2–6 min.

In addition, they found that the protein conformational changes with sonication do play a role towards the enhancement of emulsifying properties.

Heffernan et al. (2011) compared high-pressure homogenizers including a microfluidizer, an orifice nozzle homogenizer and radial diffuser homogenizer with ultrasonication (20 kHz, 1000 W) for predicting the emulsification efficiency of cream liqueur. The shelf stability was the lowest with sonication, although it was more efficient with the disruption of emulsion droplets as compared to the other configurations under similar energy density conditions. This may most likely be that cream liqueurs were over-processed at the US conditions used in the study and the interfacial membrane may have been severely damaged due to physical forces generated through acoustic cavitation. This, in turn, may provide insufficient amount of proteins available to stabilize the emulsion.

Although US is a versatile technique for making stable single dairy emulsions of desired size, there are still some instability issues in preparing double emulsions. A very recent study by Leong et al. (2017) prepared double emulsions of water-in-oil-in-water using skim milk. The study used 20 kHz US at 10 W for 40–60 s for the primary emulsification, while 2–26 W for 5 s were used for the secondary emulsification. Increased intensity of shear forces with increase in sonication power led to increased emulsion droplet disruptions (Fig. 1.1). Encapsulation efficiency increased with increasing the size of the secondary oil droplets, while a higher yield was obtained at 5 wt% loading compared to 10 and 20 wt%. The double emulsion droplets with up to 20 wt% loading were stable to phase separation for 7 days. Over-processing with US can lead to smaller droplets, but the internal phase will be leaked to the external phase (Leong et al. 2017). In addition, the use of increased ultrasonic power can also promote droplet–droplet collisions resulting in coalescence. Hence, the secondary oil droplets within the double emulsion must have a suitable size that can entrap the internal phase without any leakages, and at the same time creating a shelf-stable emulsion during storage.

Many US emulsification studies have been conducted at lab scale; however, preparation of nano-emulsions at the industrial level is still restricted by many obstacles. Industrial applications require nonstop flow, low-energy consumption, low operating and maintenance costs. One major issue associated with ultrasonic emulsification is the energy demand in producing nano-emulsions. In addition, specially designed equipment is necessary. Moreover, high-intensity US processing may generate undesired temperature increases (Abbas et al. 2013), that left unchecked, could damage the product sensory and nutritional quality. Thus, reducing the operational energy and processing cost requirements with minimal disruption to product stability and with increased functionality has emerged as a hot topic.

1.2.2 Homogenization

Homogenization is a standard dairy process practiced as a means of stabilizing the fat emulsion against gravity separation. High-pressure homogenization is frequently

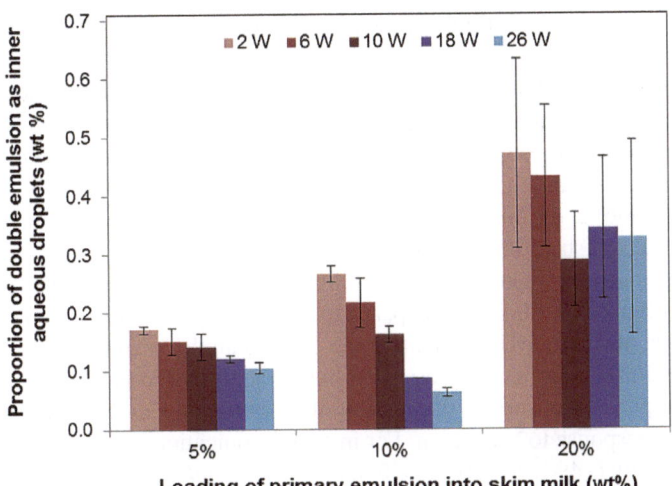

Fig. 1.1 The proportion of double emulsion as inner aqueous droplets affected by varying loading percentages (Leong et al. 2017)

utilized within the dairy industry. The size distribution and the composition of the Milk Fat Globule Membrane (MFGM) of the fat globules affect the sensorial and rheological properties of secondary dairy products (Cho et al. 1999). In recent times, ultrasound has often been used for homogenization of milk (Wu et al. 2000; Ertugay et al. 2004; Bosiljkov et al. 2011; Al-Hilphy et al. 2012). Novel dairy products with different physico-chemical and functional properties can be produced with ultrason-ication alone or in combination with other traditional homogenization techniques. The resulting differences arose from the extent of the reduction and disruption of the fat globules and MFGM, respectively. For instance, a fat globule size reduc-tion is unwelcomed in hard cheese manufacture such as Cheddar cheese, however, it provides benefits in the manufacture of soft cheeses where creamier, smoother and softer textures are desirable. Despite this, no study clearly identifies which of the homogenization techniques are suited to various dairy applications. Koh et al. (2014a) found similar particle size reductions for Whey Protein Concentrate (WPC) systems when subjected to ultrasound or high-pressure homogenization under similar energy conditions. Cavitation-derived shear forces did not result from high-pressure homogenization unlike 20 kHz ultrasonication where shear forces were derived from cavitation, highlighting the insignificant contribution of these shear effects.

The frequency, amplitude, diameter of the ultrasonic probe and experimental con-ditions such as temperature play significant roles in the physical properties of homog-enized milk. Bermúdez-Aguirre et al. (2008) found that ultrasonic homogenization (400 W, 24 kHz, using a 22 mm probe) of milk at 63 °C for 30 min decreased the milk fat globule size from 4.3 μm to <1 μm consisting a totally collapsed MFGM. Furthermore, a granular fat globule surface was obtained and was attributed to the interaction with casein micelles. Similarly, the use of continuous flow high-intensity ultrasonication (20 kHz) of milk at the power level of 150 W resulted in a fat globule size reduction of up to 82% (Villamiel and de Jong 2000). It was further found that the temperature of the treatment process influenced the fat globule size distribution. For instance, a monomodal size distribution was obtained at 70–75 °C, while a bimodal distribution resulted at lower temperatures. The extent of homogenization of milk was also found to be dependent on the amplitude (20, 60 and 100%) and the processing time (2–15 min) (Bosiljkov et al. 2011). The collapse of microbubbles induces localized shear forces that readily disrupted fat globules.

A recent study by Chandrapala et al. (2016) investigated the homogenization efficiency of ultrasound (20 kHz) compared to high-pressure homogenization (80 bar) on raw milk, ultrafiltrate (UF) retentate and cream samples at equivalent energy densities (153 J/mL). Further, they assessed the rennet and acid gelation properties of treated samples. Interestingly, the amount of fat in the system showed considerable variation in response to sonication. For instance, sonication reduced the fat globule size in raw milk (~4% w/w fat) and UF retentate (~8% w/w fat) while cream samples (~40% w/w fat) sonicated at <10 °C flocculated to form larger grapelike structures. At 50 °C, however, sonication did not form such aggregates in the same cream sample. In comparison, high-pressure homogenization at 50 °C resulted in grapelike structure formation. However, none were observed with high-pressure homogenization at low-temperature conditions. Furthermore, shorter gelation times and rennet gels with reduced syneresis were obtained with sonication. In addition, some of the grapelike structures were preserved within the coagulated gel matrix, but the larger fat globule clusters were not embedded into the matrix (Fig. 1.2). In contrast, the fat globule aggregates in the homogenized sample integrated well within the coagulated gel network. This study shows that low-temperature sonication can be adopted in lieu of high-temperature homogenization to obtain similar effects at equivalent energy demands. However, careful consideration of sonication parameters is required to avoid potentially detrimental issues that may not result from high-pressure homogenization.

1.2.3 Filtration

Membrane technology is widely adopted in the dairy industry for concentration, purification and fractionation applications due to its low-energy requirement and high versatility. However, one of the widely accepted drawback is membrane fouling reducing permeate flux. Organic and inorganic molecules are deposited on the membrane surface and in membrane pores resulting in a cake layer formation and pore plugging. Overall processing efficiency may be restored or improved with the use of complex and time-consuming chemical cleaning procedures which may cause severe membrane damage reducing membrane lifespan, and these chemical agents are potentially unsafe, expensive and create secondary pollution as chemical waste. Thus, alternative technologies are sought and US has attracted significant attention

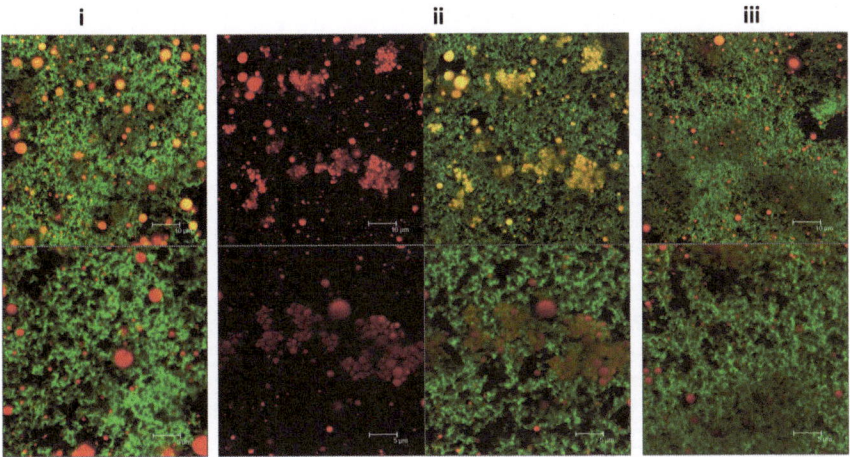

Fig. 1.2 CSLM of rennet gels produced from three different cheese–milk systems. (i) Skim milk (SM), Skim milk UF retentate (SMC) and native cream (control), (ii) SM, SMC with homogenized cream (80 bar/at 50 °C) and (iii) SM, SMC with sonicated cream (50 W/1 min at <10 °C) (Chandrapala et al. 2016)

(Muthukumaran et al. 2005a, b, 2007; Koh et al. 2014b; Gondrexon et al. 2015; Lujan-Facundo et al. 2016a, b).

Ultrasound showed a permeate flux increase during filtration by improving permeation and increasing mass transfer across the concentrated layer close to the membrane surface. Thus, the decrease of membrane surface solute concentration lowers the osmotic pressure, and in turn, increases permeation through the membrane. In addition, ultrasound provides a less compact fouling layer (Muthukumaran et al. 2005a). Micro-jets scour the surface which enhances turbulence, thereby promoting the back-transport of deposits to the bulk solution. On the other hand, the cleaning efficiency of membranes is due to ultrasonic agitation and creation of microbubbles through acoustic cavitation (Muthukumaran et al. 2005b, 2007). The energy released with the collapsing of the bubbles loosens the interactions between the foulant and the membrane, which in turn makes it easier to remove. However, the increased cleaning efficiency with ultrasound is dominated by acoustic streaming and increased turbulence as opposed to cavitation itself (Muthukumaran et al. 2005b, 2007).

Numerous other sonication studies were carried out using low-intensity US (<5 Wcm^{-2}) and low frequency (20–100 kHz) (Gondrexon et al. 2015). Figure 1.3 summarizes the US intensities and frequencies used in various membrane filtration studies.

Lujan-Facundo et al. (2016a, b) investigated the cleaning efficiency of US on Whey Protein Concentrate (WPC) solutions using four UF membranes with different molecular weight cut-offs (MWCO) (5, 15, 30 and 50 kDa), materials (ceramic and polyether sulfone) and two membrane modules (flat sheet an tubular) as a function of frequency (20, 25, 30 and 38 kHz). Low frequency was preferred and postulated to the

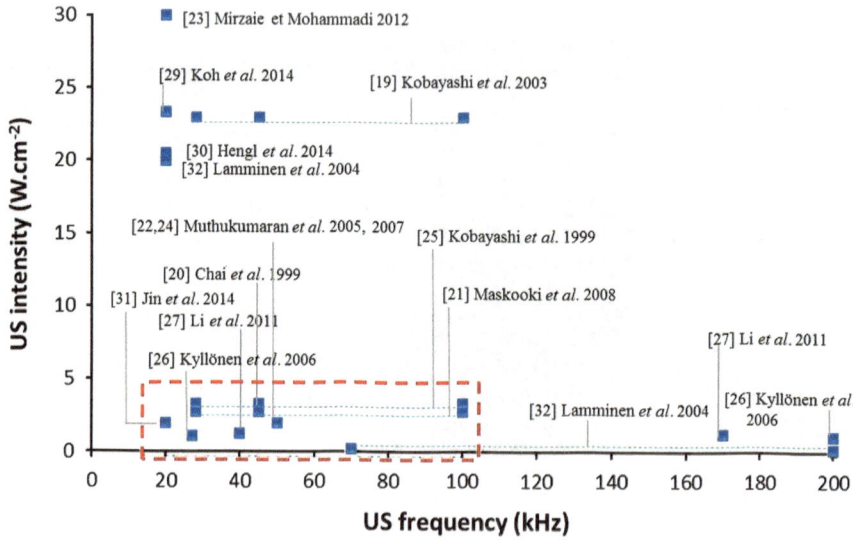

Fig. 1.3 Summary of US intensity-frequency configurations used in various membrane filtration studies (Gondrexon et al. 2015)

fact that low frequencies create large microbubbles, so their powerful collapse generates stronger vibrations which removes the foulant layer efficiently. Similarly, Caia et al. (2009) state that low-frequency and high-power ultrasound is more effective in increasing flux and improving the flux recovery due to larger cavitation bubbles that collapse more energetically separating the particles from the cake layer. The number of cavitation bubbles and an increase in acoustic energy also increases the cleaned flux ratio (Lamminen et al. 2004).

Lujan-Facundo et al. (2016a, b) further stated that US is more efficient in cleaning ceramic membranes, and the MWCO and the membrane module have little influence. Muthukumaran et al. (2005b, 2007) investigated the effects of using a surfactant towards the efficiency of ultrasonic induced UF membrane cleaning fouled by whey solutions. Interestingly, it was found that cleaning was more efficient in the absence of a surfactant, with temperature and the transmembrane pressure playing a smaller role. In addition, the frequency played a significant role towards fouling and cleaning of the membranes (Muthukumaran et al. 2007). For example, continuous low-frequency (50 kHz) ultrasound was found to be the most effective as compared to intermittent high-frequency (1 MHz) ultrasound. The less effective response by high-frequency ultrasound was attributed to a reduction in steady-state membrane flux. Increased fouling during high-frequency US treatment is likely to result from compaction of the cake layer on the membrane surface and membrane pore blockage due to lack of physical shear effects. A recent study by Jin et al. (2014) enhanced cross-flow ultrafiltration of skim milk by applying in situ ultrasonication. They found that applied US (20 kHz; 2 Wcm^{-2}) led to significant increases of permeate flux due to the disruption of the concentrated layer. However, varying US intensity from 0.6 to

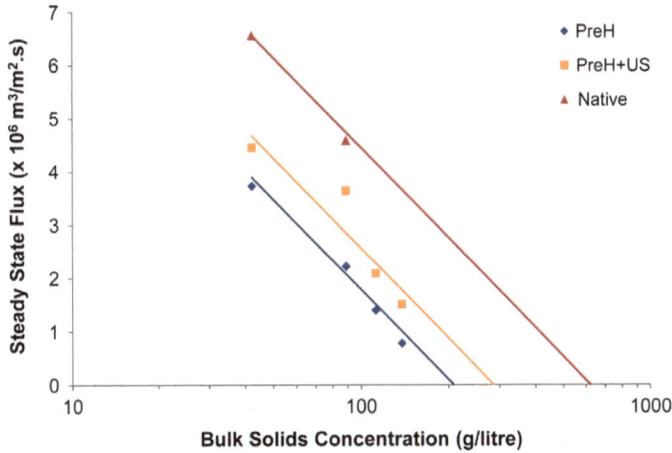

Fig. 1.4 Plot of steady-state flux obtained for native, PreH and PreH+US WPC80 feed as a function of feed concentration (Koh et al. 2014b)

2.9 Wcm^{-2} did not enhance the flux, although depended on the feed concentration which formed a highly cohesive fouling layer where it was hard to be removed by US.

Koh et al. (2014b) used ultrasound (20 kHz, 31 W) as a pretreatment step prior to a secondary heat treatment (PreH+US) to improve the UF performance of WPC solutions. Lower gel concentrations were obtained for the heat-treated (PreH) WPC due to denaturation and aggregate formation. A larger gel concentration was observed for the (PreH+US) system as compared to the PreH. This shows that sonication delays protein gelation and reduces the rate of pore blockage and cake growth (Fig. 1.4). For heat-treated feed filtration, the flux recovered by water flushing (reversible fouling) reduced considerably, while sonication increased the proportion of reversible fouling, although it was unable to reach the level observed with the native feed solution. A high proportion of foulants were removed by flushing water for preH+US sample due to the presence of smaller particles and fewer large aggregates. The flux obtained after cleaning was higher than that of the PreH feed.

Although ultrasound-assisted membrane filtration has several positive attributes, high acoustic power levels and improper installation of units can impact structural integrity of the membranes (Koh et al. 2014b). For example, Lamminen et al. (2004) observed damaged membranes during treatment at highest power (>12.2 W), while no damage was found at lower power (<7.2 W). Thus, careful consideration of US conditions must be considered in all applications and addressed individually. Ultrasonic membrane technology scale-up studies have not been reported presumably due to high energy consumption and high capital setup costs.

1.2.4 Microbial Inactivation

Milk offers a perfect setting for microbial growth, thus the microbial quality of milk and dairy products is vital. Thermal processing is the most common practice used to control the microbial growth within the dairy industry. However, various physico-chemical, nutritional and organoleptic property changes occur and due to this, alternative methods are sought. In this context, non-conventional thermal processes such as pulsed electric field (PEF), US and UV irradiation are regularly studied. These non-thermal technologies are potentially less energy-intensive, more environmentally friendly and most importantly cause minimal loss of colour, flavour and nutrition.

An antimicrobial effect created by acoustic cavitation and a bactericidal effect created by intracellular cavitation which disrupts the structural and functional components up to the point of cell lysis are mostly responsible for the ultrasound-induced inactivation of microbes (Hughes and Nyborg 1962). In addition, pressure and temperature fluctuations break down the cell walls. The free radical formation and micro-streaming disrupt and thins out the cell membranes and damages the DNA (Bermúdez-Aguirre et al. 2011). Punctured walls, breakage of plasma membrane, leakage of cytoplasmic content and damage to the inner cellular contents was also found (Alzamora et al. 2011; Guerrero et al. 2001). However, different microorganisms have varying resistance to ultrasound and depended on processing conditions.

Intense power levels and long contact times are necessary to inactivating microorganisms by ultrasound alone at ambient temperature conditions but these intense demands are vastly reduced in combination with one or more established techniques. These include combinations of US with heat (thermosonication-TS), pressure (manosonication-MS), pressure and temperature (manothermosoincation-MTS), UV irradiation (photosonication) and PEF. However, the efficiency of microbial inactivation by these methods is dependent on the treatment conditions such as amplitude of the ultrasonic waves and processing time, volume of food being processed, the type of microorganism targeted and the composition of the food. Studies have shown that thermosonication increases the lethality values of conventionally pasteurized liquid foods. This, in turn, reduces the temperature and time of the heat treatment achieving equivalent numbers of inactivated microbes and thereby improving the nutritional and organoleptic qualities (Ordóñez et al. 1987). However, the effectiveness of TS decreased with increasing temperature due to reduced cavitational effects. An increase in vapor pressure and a decrease in liquid surface tension occurs at increased temperatures reducing cavitational effects (Villamiel and de Jong 2000). The use of MS and MTS can overcome this detrimental effect.

Raso et al. (1998) investigated the effects of MS and MTS in inactivating *Bacillus subtilis* spores. MS treatment (500 kPa, 117 μm amplitude) for 12 min inactivated ~99% of the *B. subtilis* spore population, while MTS (20 kHz, 300 kPa) at 90 and 150 μm for 12 min at 70 °C inactivated ~75 and ~99.9% of the *B. subtilis* spore population, respectively. This study highlights the importance of ultrasound intensity which improves the cavitational effects through increased localized pressure and temperature areas and free radical formation regions. The use of high temperatures (>70 °C) also leads to some spore inactivation. A synergistic effect between temperature and pressure was found by Pagán et al. (1999) when applying MTS in the inactivation of *Listeria monocytogenes*. It was argued that cell envelopes that are already weakened by temperature can easily undergo changes when exposed to mechanical stress. Arroyo et al. (2011) reported that sublethally injured cells were not detected after MS at 35 °C indicating that the cytoplasmic membrane breakages resulting from MS treatment at room temperature was an all or nothing event. In contrast, damaged cells were detected when *Cronobacter sakazakii* cells were subjected to MS at 60 °C and the effect was totally attributed to heat (Arroyo et al. 2011). Table 1.1 emphasizes the efficiency of microbial inactivation in various dairy systems as affected by varying processing conditions.

The inactivation efficiency of microbes in response to ultrasound depends on the type of microorganism targeted. Gram-positive bacteria are generally more sensitive than gram-negative microbes, while spores are more resistant than vegetative cells (Halpin et al. 2013). Halpin et al. (2013) studied MTS (frequency; 20 kHz, amplitude; 27.9 mm, pressure; 225 kPa) at two temperatures (37 or 55 °C) with/without PEF (electric field strength; 32 kV/cm, pulse width; 10 ms, frequency; 320 Hz) on the inactivation of *Staphylococcus aureus, Enterobacteriaceae* and *lactic acid bacteria (LAB)*. Thermal pasteurization (72 °C, 20 s) was used for comparison purposes. On day 0, pasteurization reduced the amount of *S. aureus* in raw milk by three log cycles, while treatment with MTS/PEF at 37 and 55 °C resulted in 1.4 and 2.7 log reductions, respectively. On the other hand, *Enterobacteriaceae* on day 0 showed no significant change in milk treated with MTS/PEF at both 37 and 55 °C as compared to the control. However, on day 21, MTS/PEF treatment at 37 °C had similar levels of *S. aureus* as raw milk, while the level of *Enterobacteriaceae* was reduced with treatment. Furthermore, LAB following US treatment resulted in 50% greater log reductions as compared to conventional pasteurization (Fig. 1.5). In addition, Engin and Karagul-Yuceer (2012) found that no significant levels were found on levels of yeast and moulds in US-treated milk (20 kHz, 5 °C, 75 W, 15 min) as compared to the control. But reductions in *Staphylococcus* sp. and *Escherichia coli* were observed. Similarly, Cregenzan-Alberti et al. (2014) and Shamila-Syuhada et al. (2016) reported increased levels of inactivation of gram-negative *E. coli* and *Pseudomonas fluorescens* than gram-positive *S. aureus*. Gram-positive bacteria contains a thick and tightly adherent peptidoglycan cell wall layer which is resistant to sonication (Chemat et al. 2011). Morphology of bacterial cell may also play an important role towards varying resistances to sonication. Shamila-Syuhada et al. (2016) observed that spherical-shaped *S. aureus* was more resilient to ultrasound than rod-shaped *Lb. plantarum* and *Lb. pentosus* and attributed this behaviour to the size and surface area of the cells. Rod-shaped bacteria have a large surface area, while spherical-shaped microbes are of smaller sizes (Chemat et al. 2011).

It is well known that ultrasound in combination with heat has produced positive outcomes in inactivation of microorganisms. However, the composition of the food system influences the rate of inactivation. Bermúdez-Aguirre and Barbosa-Cánovas (2008) studied the effect of butterfat content (0–3% w/w) in milk in the inactivation

Table 1.1 Some literature data on the inactivation of microbes in various milk systems with different US processing conditions

Food matrix	Target microbes	Treatment conditions	Findings	Reference
Milk with 4% fat	*Staphylococcus aureus* *Escherichia coli*	20 kHz Temperature (20, 40 and 60 °C) Amplitude (60, 90 and 120 µm) Time (6, 9 and 12 min)	*E. coli* is more susceptible for inactivation than *S. aureus* Optimum conditions: Temperature = ~60 °C, Time = 12 min, Amplitude = 117.3 µm to *S. aureus* and 110.4 µm to *E. coli* Higher amplitude, longer processing time and high temperature increased the inactivation	Herceg et al. (2012a)
Raw bovine milk with 4% fat	*Mesophillic bacteria*	20 kHz (12 mm diameter) Amplitude (60, 90 and 120 µm) Temperature (20, 40 and 60 °C) Time (6, 9 and 12 min)	Inactivation increased with increase in amplitude, time and temperature Optimum conditions: T = 45.34 °C, Time = 9.84 min, Amplitude = 120 µm	Herceg et al. (2012b)
Raw bovine milk	*Enterobacteriae*	20 k Hz Amplitude (60, 90 and 120 µm) Temperature (20, 40 and 60 °C) Time (6, 9 and 12 min)	Inactivation increased with increase in amplitude, time and temperature	Juraga et al. (2011)
Raw milk	*L. monocytogens* Aerobic bacteria	Thermosonication (20 kHz) at 57 °C for 18 min	5 log reduction of *L. monocytogens* and total aerobic bacteria	D'Amico et al. (2006)
Non-fat, low-fat and full fat milk	*L. monocytogenes*	Multifrequency Dynashock US	Inactivation was biphasic Slowest log-linear inactivation rate of −0.24 log cfu/min in full-cream milk Fastest inactivation rate of −0.37 log cfu/min in low-fat milk	Gabriel (2015)

(continued)

Table 1.1 (continued)

Food matrix	Target microbes	Treatment conditions	Findings	Reference
Raw whole milk	*L. innocua* *Mesophillic bacteria*	Thermosonication (24 kHz) Amplitude (0, 40, 72, 108 µm) 63 °C Up to 30 min	Thermal pasteurization obtained 0.69 log reduction. Thermosonication 5 log reduction at 60, 90 and 100% amplitude. Both species behaved similarly. No change to acidity or pH of the systems. Whiter colour achieved	Bermúdez-Aguirrre and Barbosa-Cánovas (2008)
Raw whole milk	*Mesophillic*	Thermosonication (24 kHz, 63 °C) Amplitude of 108 µm. Treatment time of 30 min. As a function of storage time	Reduced from 4.7 log cfu/ml to 2 log cfu/ml day 0. <1 log cfu/ml on day 16	Bermúdez-Aguirrre and Barbosa-Cánovas (2008)
Rehydrated powdered milk	*Cronobacter sakazakii*	Manosonication and Manothermosonication (20 kHz, 35–68 °C at constant pressure 200 kPa) 117 µm amplitude, 113 W	Higher inactivation with MTS as compared to heating alone in between 50 and 68 °C. Inactivation effect of MTS at 68 °C was equivalent to thermal treatment 4 min, 35 °C, 200 kPa, 117 µm—99.99% death of *C. sakazakii* cells	Arroyo et al. (2011)
Skim milk powder	*Geobacillus stearothermophilus vegetative sells and spores*	ThermosonicationTotal solids—9–50% Temperature 55–75 °C Amplitude—240 um, 20 kHz	Optimized conditions cell reduction. Inactivation was most effective before 9.2%, 75 °C, 10 s and after 50% TS, 60 °C 10 s for both cells and spores	Beatty and Walsh (2016)
Skim milk	*Bacillus cereus*	Thermosonication 1.5 min, 70 °C, HIU	Log reduction <0.5	Evelyn and Silva (2015)

(continued)

Table 1.1 (continued)

Food matrix	Target microbes	Treatment conditions	Findings	Reference
Raw bovine milk	*Escherichia coli* *Pseudomonas fluorescens*	Temperature (20–52 °C) Intensity (60–120 Wcm^{-2}) Treatment time (40–240 s) Pressure 225 kPa	Reductions of up to c. 1.6 log CFU/ml were achieved for *Escherichia coli* and *Pseudomonas fluorescens* at 36 °C, 90 Wcm^{-2}, 240 s. Lower inactivation values were reported for *Staphylococcus aureus*	Cregenzan-Alberti et al. (2014)
Raw milk	*P. fluorescens*	Thermosonication 51 °C, 20 kHz, 150 W 102.3 s	An inactivation of 1.1 log cfu/ml Increased with time and amplitude	Villamiel and de Jong (2000)
UHT milk	*Staph. aureus* *L. monocytogenes* *Lb. plantarum* *Lb. pentosus.* *S. Typhimurium* *E. coli* *P. fluorescens*	62.5 and 125 μm Time 5–15 min 24 kHz	The rate of inactivation increased with sonication time at both amplitudes Inactivation was greater at 125 μm amplitude than at 62.5 μm *Staph. aureus* and *L. monocytogenes* were more resistant to sonication than *Lb. plantarum* and *Lb. pentosus* *S. Typhimurium, E. coli* and *P. fluorescens* were equally susceptible to sonication	Shamila-Syuhada et al. (2016)
Milk	*E. coli* *L. monocytogenes* *P. fluorescens*	20 kHz, 750 W 124 μm amplitude	*E. coli* and *L. monocytogenes* counts were reduced by 100 and 99%, respectively after 10 min *P. fluorescens* counts were completely eliminated after 6 min	Cameron et al. (2009)

(continued)

Table 1.1 (continued)

Food matrix	Target microbes	Treatment conditions	Findings	Reference
Whole milk Skim milk SMUF Phosphate buffer	E. coli L. monocytogenes	85 W/cm^2 24 kHz, 100 μm amplitude, pulsing 80%	Inactivation of >90% of E. coli and L. monocytogenes was obtained at 2.4, 2.4, 2.2 and 9.3, 8.6, 7.6 min, respectively with whole, skim and phosphate buffer Lactose presence increased inactivation due to the stabilization effect of lactose of the bacterial membrane and proteins or accumulation of compatible solutes	Gera and Doores (2011)
Milk		Thermosonication 400 W, 24 kHz, 63 °C 30 min Energy delivered: 129 mW/ml	Increased shelf life A viable option for pasteurization	Bermúdez-Aguirre et al. (2009a)
Raw whole milk	L. innocua Mesophillic bacteria	Thermosonication (400 W, 24 kHz)	Pasteurization reduced 0.69 and 5.5 log after 10 and 30 min, respectively, while with US for 10 min 5 log reduction was attained Improved whiteness	Bermúdez-Aguirre et al. (2009b)
Low-fat UHT milk	L. innocua	TS + Pulse electric field Preheat = 55 °C 24 kHz 400 W PEF 40 kV cm^{-1}	Similar cell death numbers were achieved for US and pasteurization treatments	Noci et al. (2009)

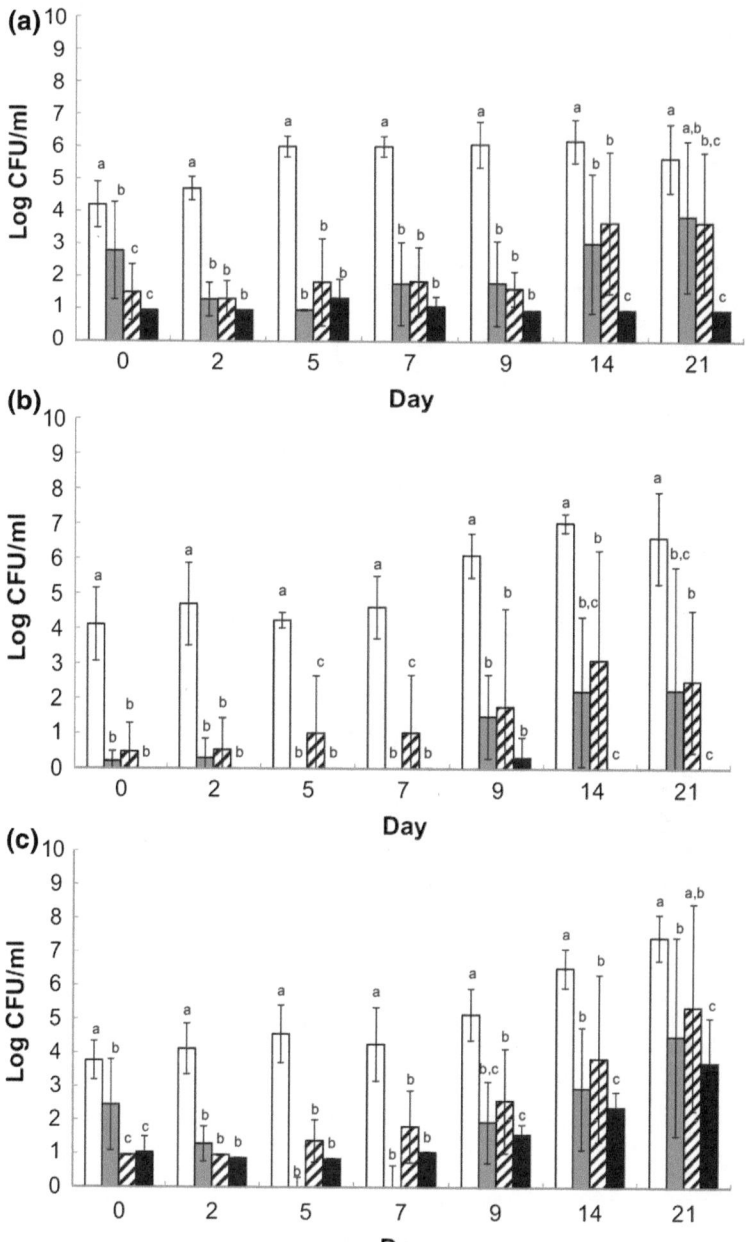

Fig. 1.5 **a** *S. aureus*, **b** *Enterobacteiraceae*, **c** LAB as a function of storage time for raw milk (white bar), MTS/PEF at 37 °C (grey), MTS/PEF at 55 °C (stripped) and thermally treated milk (black) (Halpin et al. 2013)

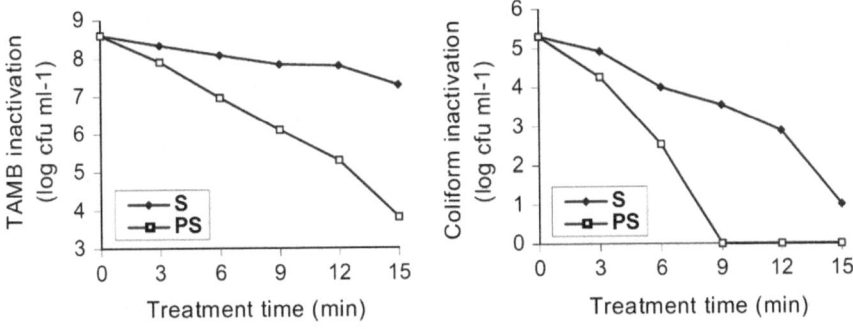

Fig. 1.6 Inactivation of total mesophillic and coliform bacteria in milk as a function of treatment time—Sonication (S) and photosonication (PS) (Sengul et al. 2011)

of *Listeria innocua* at 63 °C for 30 min, sonication at 24 kHz and 120 μm amplitude. It was found that the rate of inactivation decreased with increasing fat content by creating a protective effect and the cells not receiving the equivalent physical effects. US disrupts the MFGM generating smaller sized fat globules with a new rough surface (Bermúdez- Aguirre and Barbosa-Cánovas 2008). The microorganisms may conceal within the rough fat globule surface and be protected from the effects of acoustic cavitation; while no protective effect was found in fat-free milk systems. In addition, the changes in US energy distribution (Earnshaw et al. 1998), intensity of cavitation and the boiling point of some soluble components present in full fat and free fat milk systems may influence the inactivation efficiency (Carcel et al. 2009).

Combination of two or more non-thermal processes achieves the desired effect at reduced individual treatment intensities (Ross et al. 2003). For instance, reduced treatment times and lower energy consumption was achieved with photosonication (US +UV light) as opposed to US and UV irradiation alone. Sengul et al. (2011) studied the effect of photosonication in reducing the total and coliform bacteria in raw milk. The total and coliform bacteria reduction achieved was 4.79 and 5.31 log cfu mL^{-1}, respectively with photosonication (120 mm; 240 W 24 kHz US; UV 13.2 Wcm^{-2}; 15 min) while 1.31 and 4.01 log cfu mL^{-1} reduction was reported for total and coliform bacteria, respectively with sonication alone (Fig. 1.6). Ultrasound enhances the efficiency of UV light penetration with UV directly damaging DNA (Engin & Yuceer, 2012).

Although inactivation of microbes shows promise, sensory attributes are also important. Marchesini et al. (2015) studied different amplitude conditions (70 and 100%) and time variations (50–300 s) on inactivation of *E. coli, P. fluorescens, S. aureus* and *Debaryomyces hansenii* in milk. Application of US at 100% amplitude conditions for 300 s led to a population reduction of 4.61, 2.75, 2.09 and 0.55 log for *D. hansenii, P. fluorescens, E. coli* and *S. aureus*, respectively. But the milk sensorial attributes were changed significantly resulting in metallic, burnt, rubbery and sharp flavours. Volatile compounds described as 'off flavours' are proportional to energy density and may be controlled by selecting appropriate treatment conditions as demonstrated in other studies (Riener et al. 2009; Martini and Walsh 2012; Juliano et al. 2014). In addition, some microorganisms believed to be "killed" may only be sublethally injured. Injured cells pose a threat to food integrity as they are unpredictable and have the potential to become viable under favourable environmental conditions. It is believed that after treatments, by either thermal or non-thermal technologies, there may be one population of microbes which are dead, another population that are viable, and a third population that are sublethally injured (Wu 2008). It is of the utmost importance to be able to distinguish between viable cells and impaired cells in order to gain complete food safety (Wu 2008). Researchers have reported that improvements in the design and efficiency of US chambers and US horn geometries lead to more efficient microbial inactivation. However, specialized equipment which can deliver higher acoustic intensity and power is required. In addition, US chambers and horn geometries tailored for specific applications are required to achieve certain targeted effect. Such issues must be addressed for ultrasound to have a future in large-scale operations.

1.2.5   Functionality Modifications

Ultrasound has been used in recent years to improve processing effectiveness and to manufacture dairy products with 'tailored' functionality by altering microstructure through component interactions (Knorr et al. 2004).

1.2.5.1   Solubility of Dairy Powders

Rapid solubilisation of dairy powders is desired by the industry to maintain the quality of secondary dairy products and avoid high production costs and prolonged processing times. Chandrapala et al. (2014a, b) investigated the solubility of milk protein concentrate and micellar casein powders as affected by ultrasonication at the energy density of 153 J/ml and compared these to high-pressure homogenization, high-shear rotor–stator mixing and low-shear overhead stirring. The high shear techniques greatly accelerated the solubilization of these powders. The shear forces broke apart the powder agglomerates and accelerated the release of individual casein micelles into the solution without affecting the structure of the individual solubilized proteins. The localized high shear forces exerted by acoustic cavitation did not significantly affect the mass transfer of minerals from the casein micelles when the mineral balance between the casein micellar and serum phases were analysed after sonication (Chandrapala et al. 2014a, b). Only the large powder particles were affected (Fig. 1.7). Another study by McCarthy et al. (2014) showed that ultrasound (20 kHz/70.2 W) significantly increased the solubilization of MPC powders. However, their study used a pre-stirring step at 50 °C, which may have aided the dissolution to a certain extent. Further, the study did not use controlled temperature conditions and the sam-

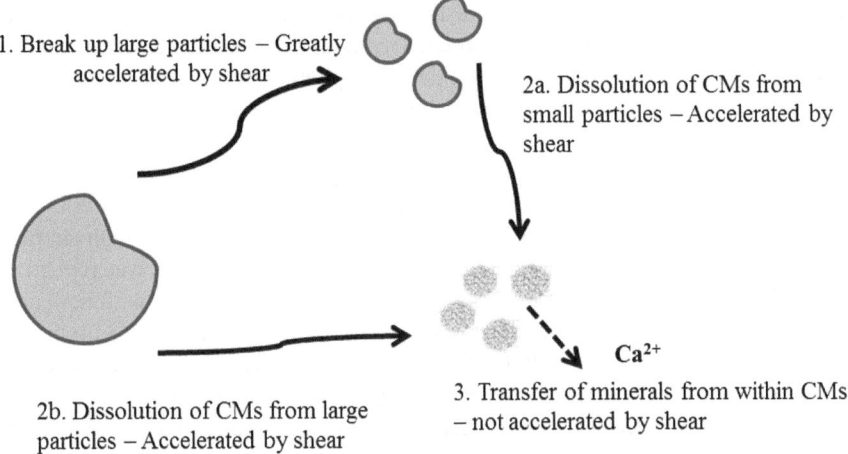

1. Break up large particles – Greatly accelerated by shear

2a. Dissolution of CMs from small particles – Accelerated by shear

2b. Dissolution of CMs from large particles – Accelerated by shear

Ca²⁺

3. Transfer of minerals from within CMs – not accelerated by shear

Fig. 1.7 Schematic representation of the effect of shear forces on the solubilisation of dairy powders (Chandrapala et al. 2014a)

ple temperature increased to 70 °C during sonication resulting in denaturation and aggregation of proteins which may have also affected the solubility.

Furthermore, solubility of dairy powders declines with storage (Fyfe et al. 2010). During storage, increased protein interactions occur leading to compaction and the formation of an aggregated skin of proteins. Chandrapala et al. (2014b) investigated the effect of storage on the solubility of milk protein concentrate, calcium caseinate and whey protein concentrate powders that were sonicated prior to lab-scale spray drying. Two different humid conditions (22 and 75%) at 25 °C for 60 days were evaluated. The microstructural studies showed the presence of particle agglomerates in non-sonicated powder samples upon storage, whereas none were visible in the sonicated samples. However, no substantial compositional changes to the surface in terms of fat, protein and lactose contents were observed. The study showed that sonication prior to drying increased the powder stability during storage and delayed the loss of powder solubility with storage. Similar results were obtained by Yanjun et al. (2014) where they investigated the effect of US (12.5 W, 0.5–5 min) prior spray drying of milk protein concentrates and showed significant increases in powder solubility.

O'Sullivan et al. (2014, 2015b) showed particle size reductions in sodium caseinate and MPI sonicated at US 34 Wcm⁻² for 1 min followed by no change on prolonged sonication. They hypothesized that this size reduction is due to the disruption of protein micelles along with changes to the electrostatic and hydrophobic interactions by high shear forces generated through acoustic cavitation. However, they further found that after 7 days, the particle size increased with sodium caseinate, most probably due to the reorganization of the proteins into smaller sub associates with non-covalent attractions such as hydrophobic and electrostatic. In contrast, a further size reduction was observed with MPI upon storage indicating further continuation of size reduction due to acoustic cavitation and shear forces. Furtado et al. (2017) found no MW changes of the proteins with sonicated sodium caseinate (300 W, 2–6 min) although sonication exposed the hydrophobic groups with an increase in surface hydrophobicity and no change in charge. Similarly, circular dichroism spectra revealed no secondary protein structural changes for sodium caseinate. Further, intrinsic viscosity was positively affected by the increase in sonication time.

In recent studies, sonication-induced structural changes of casein micelles have been investigated due to the possibility of alterations due to extremely high temperatures and pressures by acoustic cavitation. Madadlou et al. (2009) showed reductions of the size of reassembled casein micelles which were structurally and functionally different to native casein micelles after exposure to ultrasound (35 kHz frequency) for 6 h at pH >8. Another study by Nguyen and Anema (2010) showed a decrease in size of the native casein micelles with sonication (22.5 kHz; 50 W). Contrary to these studies, Chandrapala et al. (2012a) showed that sonication merely reduces the size of the fat globules and whey protein aggregates, but did not affect the casein micelles, their composition nor the mineral balance in fresh skim milk (Chandrapala et al. 2012a). Similarly, no changes to the casein micelles and partial disruption of whey proteins from whey–whey aggregates were reported by Shanmugam et al. (2012) following sonication (20 kHz, 20/40 W). Contrary to these studies, Liu et al. (2014a) found the disruption of the casein micelles but only at high pH values (6.7–8) after sonication (286 kJ/kg/20 kHz). However, the dissociation of κ-casein was found to be dominated by the increase in pH (Anema and Klostermeyer 1997). Thus, the release of individual proteins from casein micelles at high pHs under the influence of sonication is arguable and unlikely to be an ultrasound effect alone but a combination of treatment and environment.

1.2.5.2   Foaming

Foams are gas–liquid systems. Foam properties depend on the physico-chemical characteristics of the phases, the production method and processing conditions. Jambrak et al. (2008) showed that high-intensity low-frequency ultrasound (20 kHz) had a significant effect on the foaming ability of whey proteins. The efficient homogenization effect by low-frequency ultrasound may improve the foaming property. During sonication, protein structure unfolds due to physical shear and temperature, and thus increases the foaming ability (Jambrak et al. 2008, 2010). Contrarily, the use of high-frequency ultrasound (500 kHz) did not impact the foaming ability of whey protein, although it affected the solubility and conductivity. This may be mainly due to the reduced physical effects at higher frequencies. In addition, they showed that sonication at 40 kHz using a bath reduced the foaming ability of whey proteins compared to 20 kHz. It was argued that the direct contact between the horn tip and the sample was favoured while the ultrasonic bath had no direct contact with the irradiating surface.

Tan et al. (2015) studied the effect of ultrasonic energy (20–60% amplitude) and processing time (5–25 min) on the foaming ability of 15 wt% WPC dispersions as a function of concentration (10–20 wt%). The highest foam overrun which is indicative of greater foaming capacity was achieved when the highest amplitude of 60% and the longest time of 25 min were applied. Shear forces break protein agglomerates, unfolding and exposing the charged and hydrophobic and hydrophilic segments of the amino acid chains. This newly formed unfolded protein structure with reduced surface tension at the air–water interface enables efficient trapping of air resulting in good foaming properties (Venugopal 2008). In addition, the increased surface activity near the interface helps stabilize the WPC foams with increased flexibility following sonication (Weiss et al. 2011). They further found that foam overrun was significantly affected by the protein concentration, sonication time and ultrasound amplitude. A better foam texture was obtained as the protein concentration increased due to greater suspension viscosity while a higher stability was achieved at higher amplitude due to the strengthening of protein films at the interface of the foam slowing the rate of foam collapse (Weiss et al. 2011). High surface viscosity forms strong foams that are more capable of compression and expansion (Kinsella 1981).

1.2.5.3   Heat Stability

Heating whey proteins at temperatures ≥ 70 °C causes denaturation, followed by aggregation (Chandrapala et al. 2011). Heat treatment is unavoidable during the processing of dairy products and excessive thickening or gelling during heating can occur. Ashokkumar et al. (2009a, b) and Zisu et al. (2010) established an approach to overcome the heat stability issues by combining ultrasound with a heat treatment step to break the heat-induced whey protein aggregates and prevents their reformation on further heating. This, in turn, prevented increase in viscosity and gelation (Fig. 1.8a). Similarly, Yanjun et al. (2014) showed significant decrease in viscosity by applying an ultrasound treatment after a preheating step and was attributed to the decrease in particle size. Initially, it was claimed that these observed viscosity changes after sonication were due to a combination of both physical and chemical effects created by acoustic cavitation. However, the amount of free radicals generated at 20 kHz frequency is low (Devi et al. 2005) and with the greatest viscosity reductions occurring at this frequency, it is unlikely that chemical reactions play a major role.

Chandrapala et al. (2012b) conducted a detailed investigation of the protein aggregation mechanism with and without an ultrasound step prior secondary heating. The surface charge which is indicative of electrostatic interactions and reactive thiol groups, which are indicative of thiol–disulfide interactions remained unchanged after sonication in between the preheating and post-heating steps (Fig. 1.8b). However, the surface hydrophobicity of these aggregates which is indicative of hydrophobic interactions was altered (Fig. 1.8b). It was concluded that physical forces generated through acoustic cavitation broke down the protein aggregates, leading to the formation of smaller aggregates with a lower surface hydrophobicity. The newly formed smaller aggregates with low surface hydrophobicities are resistant towards further

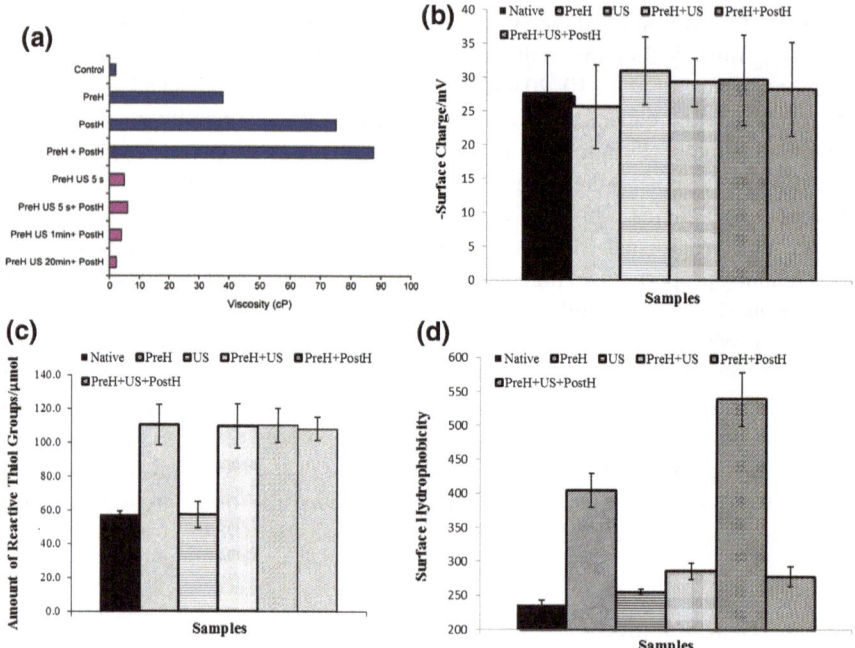

Fig. 1.8 **a** The viscosity changes, **b** Surface charge, **c** reactive thiol content and **d** surface hydrophobicity of 6.4% reconstituted whey protein concentrate solutions under various processing conditions: control (C), preheating (PreH), sonication (US-20 kHz, 31 W), and post-heating (PostH) (Adapted from Ashokkumar et al. 2009a, b; Chandrapala et al. 2012b)

aggregation during post-heating. This, in turn, improves the heat stability. Shen et al. (2017) investigated the physico-chemical properties of thermally aggregated whey proteins as affected by high-intensity ultrasound (20 kHz, 31 Wcm^{-2}) prior to and after a preheat treatment (85 °C/30 min) process. They showed a significant increase in surface hydrophobicity and free thiols without affecting the total thiol content. They hypothesized that the disruption of more non-covalent interactions as opposed to disulphide bonds was accountable for the observed effects.

Yanjun et al. (2014) suggested that the sonication-induced changes of whey protein solutions at a frequency of 20 kHz were mostly macromolecular level physical modifications without protein structural changes. Similarly, Frydenberg et al. (2016a, b) investigated the effect of thermosonication (24 kHz, 300 Wcm^{-2}) on the secondary structure of WPI samples with varying α-LA and β-LG ratios as a function of pH. Thermosonication-reduced enthalpy of whey protein denaturation with greater reductions at higher pH. The protein structure loosens when pH moves away from the isoelectric point as the surface charge density increases and repulsive forces are higher. However, no secondary structural changes were observed. They hypothesized that the high pressure generated by acoustic cavitation induces an open protein state where water easily permeates. This, in turn, replaces the intramolecular hydrogen bonds with water–protein hydrogen bonds and the strong water–protein hydrogen bonds then lead to an intact secondary protein structure. Similarly, Chandrapala et al. (2011) showed no substantial protein structural changes with up to 60 min sonication of WPC solutions at 31 W. In contrast, Kresic et al. (2008) investigated the rheological and thermophysical properties of WPC and WPI solutions subjected to sonication (20 kHz, 15 min). They attributed the observed flow behaviour changes to an altered protein structure. It was hypothesized that the hydrophilic amino acid segments of the proteins are exposed with sonication leading to an increased binding of water molecules. Similarly, Stathopulos et al. (2004) showed amyloid fibril formation with sonication (20 kHz, 30 W) of BSA protein solutions. However, this study used excessively high ultrasonic energy and small sample volumes.

Martini and Walsh (2012) treated whey solutions containing 28.2% total solids with US (15 W, 5 min, 60 °C) and investigated the thermal stability and sensory evaluation. Interestingly, no significant changes in sensory attributes were observed. Moreover, several unfavourable attributes such as cardboard and malty were less prominent in sonicated samples.

1.2.5.4 Heat-Induced Gelation

Whey protein isolates and concentrates are widely used as ingredients due to their highly desirable functional and nutritional qualities. Heat-induced whey protein gels provide a desirable texture and improved water holding capacities to various food systems. Zisu et al. (2011) showed increased gel strengths along with reduced gelation times and syneresis after sonication (20 kHz, 13–50 W, 1–60 min) of WPC suspensions. The microstructure showed compact densely packed whey proteins. These improvements were observed across the pH range of 6.7 and 9.5 and it was suggested that increased gelation characteristics exerted by ultrasound were different to the effects of pH. Further, they showed no such effect with WPI solutions and were attributed to the absence of large protein aggregates in the original solution. Frydenberg et al. (2016a, b) studied the effects of US (24 kHz, 300 Wcm^{-2}) on the heat-induced gelation characteristics of WPI as a function of varying α-LA and β-LG at 65 °C or at 20 °C. High α-LA content increased the gel strength under thermosonication conditions and it was argued that heat and US together initiated extensive unfolding of α-LA where four intramolecular disulfide bonds are disrupted, rendering α-LA in forming eight disulfide bonds with α-LA and β-LG, thus form strong gels. A denser and more compact gel network also led to reduced water holding capacity.

The composition of the food system plays a significant role towards the functionality changes initiated by US. Saffon et al. (2011) investigated the impact of the presence of buttermilk concentrate towards the denaturation kinetics with US treatment at 275 W using a 20 kHz horn. Ultrasound treatment increased the proteins aggregation yield and decreased the water holding capacity (WHC) and this effect was dependant on the proportion of buttermilk in the mixtures.

1.2.5.5 Fermentation

Fermentation increases the shelf life of a dairy product while enhancing the taste and improving digestibility. However, in some instances, long processing times and insufficient fermentation are unavoidable. Wang and Sakakibara (1997) investigated milk fermentation using different Lactobacillus strains, including *Lactobacillus delbrueckii* subsp., *bulgaricus B-5b, Lactobacillus helveticus LH-17, L. delbrueckii* subsp., *lactis SBT-2080* and *Lactobacillus acidophilus SBT-2068* following son-ication (17.2 kWm^{-2}). The viable cell count decreased with sonication compared to non-US cell counts. Sonication-induced the release of galactosidase from lactic acid bacteria cells increasing the galactosidase activity, although the hydrolysis of lactose was dependant on the effective release of galactosidase. Another study by Nguyen et al. (2012) investigated the effect of high-intensity ultrasound (20 kHz; 100 W 7–30 min) on carbohydrate metabolism in milk fermentation by *Bifidobac-terium breve ATCC 15700, Bifidobacterium infantis, Bifidobacterium animalis* subsp., *Bifidobacterium lactis (BB-12)*, and *Bifidobacterium longum (BB-46)*. The hydrolysis of lactose and the trans-galactosylation reaction in all sonicated milk were enhanced during 24 h of fermentation. Ultrasonication affected the production of organic acids during fermentation depending on the type of bacteria used. For instance, the production of major organic acids increased with *BB-12, B. breve*, and *B. infantis*, while the ratio of acetic acid to lactic acid and the ratio of total acetic and propionic acids to lactic acid decreased in *BB-12* and *BB-46* samples, respectively. The study concluded that high-intensity ultrasound is a potential alternative to traditional processing without the need for additional external prebiotics and galac-tosidase. Barukcic et al. (2015) studied the effect of ultrasound on fermentation of thermosonicated whey. Different power levels were used (480–600 W) for 6.5, 8 and 10 min at temperatures of 45 and 55 °C. Increased membrane permeability from the cell wall of dairy cultures treated with ultrasound allows the release of intracellular enzyme β-galactosidase and is responsible for the improved fermentation processes. Sonication of milk at 480 W, 55 °C for 8 min resulted in a notable increase in viable cell counts at the end of the fermentation with better sensory properties indicating that US can be used to modify product quality, and also to improve the production of microbial biomass using whey as a substrate.

1.2.5.6 Rennet Gelation

The physico-chemical changes induced by sonication may influence the rennet gela-tion properties depending on the severity of the processing conditions and the com-position of the milk system as discussed previously. Several studies were performed under different experimental conditions in recent years. Liu et al. (2014b), observed the renneting properties of milk treated with ultrasound (20 kHz, 286 kJ/kg) as a func-tion of pH (6.7, 8). Firm gels were observed with sonicated milk as opposed to non-sonicated milk. The gelation characteristics such as gelation time and texture were significantly modified in rennet gels made from milk sonicated at pH 8.0 and readjusted back to pH 6.7 as compared to gels made from milk sonicated at pH 6.7. The modified renneting behaviour was attributed to ultrasound-induced protein structural changes in milk (Liu et al. 2014b). However, many studies found no changes to the structure of casein micelles with sonication (Chandrapala et al. 2012a; Shanmugam et al. 2012). Furthermore, Chandrapala et al. (2013) evaluated the phosphate-induced micellar casein gel characteristics as affected by sonication (20 kHz). Sonication of 5 wt% micellar casein (MC) solution before the addition of 7.6 mM Tetra Sodium Pyro Phosphate (TSPP) formed a firm gel with a fine protein network and low syneresis. In contrast, sonication after the addition of TSPP led to an inconsistent weak-gel structure with high syneresis. The results indicated that the state of the casein micelles prior sonication significantly influenced the gelation characteristics of phosphate-induced micellar casein gels and should be carefully controlled. Thus, the increased rennet gelation characteristics observed by Liu et al. (2014b) are not solely responsible for ultrasound but is a combined effect of pH and ultrasound.

The physico-chemical properties of milk strongly affect cheese yield and the coagulation properties. However, goat milk coagulation is associated with weaker gels and greater whey separation. These poor coagulation properties lead to more protein losses, lower cheese yields and reduced textural integrity. The effects of US pretreatment (800 W, 0–20 min) of goat milk prior rennet-induced coagulation was studied by Zhao et al. (2014) to evaluate the improvements of milk coagulation properties. US treatment increased the extent of whey protein denaturation and soluble calcium and phosphorus contents by 9.6, 16.9 and 13.7%, respectively. In addition, the gelation characteristics such as firmness, strength, cohesiveness and water holding capacity were increased. Furthermore, US-treated gels showed highly branched honeycomb-like structures with many small uniform pores contrary to the larger pore sizes and fewer interconnections without sonication.

1.2.5.7 Age Gelation

Age thickening is well known within the dairy industry where viscosity of concentrated solutions increases with storage under low shear. This is undesirable especially in the manufacture of spray dried dairy powders. High-power low-frequency ultrasound (20 kHz) was found to lower the viscosity of concentrated milk prior to spray drying. A study by Zisu et al. (2012) showed that batch sonication for 1 min at 40–80 W and delivering an energy density of 4–7 JmL^{-1} reduced the viscosity of medium-heat skim milk concentrates (50–60% solids) by 10%. The viscosity reduced by another 7% with further increase in the solids concentration of the concentrated milk, however, sonication only delayed the rate of thickening for milk systems that have already started ageing. In addition, sonication showed changes in the shear thinning behaviour at shear rates below 150 s^{-1}.

1.2.5.8   Acid Gelation

The nature of the acid gel plays a significant role towards the quality of yoghurt. A smooth texture with low syneresis is highly desired by the consumer. During acidification, loss of steric stabilization of casein micelles and increasing tendency of denatured whey proteins to be associated with casein micelles prevails. The resultant aggregates of whey proteins and caseins act as bridging materials due to the reduced repulsive charge forming the yoghurt gel network (Horne and Davidson 2003). Denaturation of whey proteins with ultrasonication leads to the formation of whey–whey and whey–casein aggregates through disulfide-mediated covalent bonds (Shanmugam et al. 2012). This additional denaturation of whey proteins strengthens the gel matrix (Shanmugam et al. 2012). A recent study by Nobel et al. (2016) stated that the US effects on acid gelation varied with pH. They proposed three fermentation stages. At pH >5.4, the casein micelles were stabilized by electrostatic interactions, so that only temporary and reversible interactions were modified by US. At pH between 5.4 and 5.1, non-reversible formation of large protein particles was possible with sonication. At pH <5.1, a homogeneous gel network stabilized by casein micelles existed and the particles were too large to be affected by ultrasound. Some of the literature discussing the effects of ultrasound on acid gelation is highlighted in Table 1.2.

Water molecules homolyzed by cavitation generates highly reactive free radicals. These homolyzed water molecules can then react with the milk proteins and fat globules (Vercet et al. 2002). In addition, physical forces generated through acoustic cavitation can disturb large macromolecules or protein particles. Vercet et al. (2002) suggested that the alterations to the size of the fat globules did not play a significant role towards the increased gelation kinetics, although a thorough investigation was not performed to investigate the effects of fat globules towards the gelation kinetics. A recent study by Nguyen and Anema (2017) investigated ultrasonication of whole milk at 22.5 kHz and 50 W on the acid gelation properties. Acidification of US milk-produced gels with increased firmness and reduced gelation times compared to untreated milk. In skim milk, <40% of whey proteins were denatured after 10 min ultrasonication and approximately 80% were denatured after 30 min ultrasonication (Nguyen and Anema 2010),. In whole milk, the levels were 80 and 100% at equivalent processing times, respectively. The proteins present in the MFGM denature at low temperatures and promote the denaturation of the whey proteins (Ye et al. 2004). Ultrasonication causes considerable homogenization of the fat globules. The surface area of the fat globules increased by ~50% with sonication. The newly exposed fat globule surface may encourage the unfolded whey proteins to interact with the fat globules, and increase the level of irreversibly denatured whey proteins in comparison to similarly treated skim milk samples. Furthermore, increased temperature along with sonication of whole milk showed increased gel strengths (Nguyen and Anema 2017). This is in contrast with acid gels made with skim milk (Nguyen and Anema 2010). This again indicates that the effects of separate heat and sonication on acid gelation are specific to whole milk containing fat globules.

Table 1.2 Some literature data on acid gelation properties as affected by US and combination of techniques

Variables	Comments	Reference
Thermosonication 25 kHz, 400 W, 45/75 °C, 10 min	Higher gelation pH values Increased viscosity Higher WHC Honeycomb gel structure with a porous nature averaging a structural size of 2 μm	Riener et al. (2009)
Thermosonication 24 kHz, 45 °C, 10 min	Increased gel strength Increased WHC Decreased syneresis Sensory: superior texture and colour properties Increased with increase in fat content from 0.1 to 1.5%	Riener et al. (2010)
Manothermosonication 40 °C, 20 kHz, 12 s, 2 MPa	Increased flow curves, viscosity, yield stress, viscoelastic properties Rigid structures Increased rheological properties	Vercet et al. (2002)
Ultrasonication 20 kHz, 150–750 W, 10 min	Increased WHC and viscosity Decreased syneresis	Sfakianakis and Tzia (2010)
Ultrasonication 20 kHz, 50–500 W, 1–10 min	Increased WHC and viscosity Decreased syneresis Increased effects with increase in amplitude and time	Wu et al. (2001)
Ultrasonication 20 kHz, 150–750 W High pressure homogenisation 2 stages (20–30 MPa/5 MPa)	Decreased the pH lag phase duration Increased the pH reduction rate Increased viscosity Effects were more positive than HPH	Sfakianakis et al. (2015)
Ultrasonication 450 W, 150 mL, 3000 kW/m^3, 8 min	Decreased fermentation time from 207 to 176 min No change for samples sonicated prior inoculation	Nobel et al. (2016)

Shanmugam and Ashokkumar (2014b) studied the acid gelation properties of 7% flaxseed oil: milk emulsions treated with 20 kHz ultrasound (US) at 176 W for 1–8 min. The sonication process improved the gelation characteristics such as decreased gelation time, increased elastic nature, decreased syneresis, increased gel strength and increased shelf life. The presence of finer emulsified oil globules which were stabilized by the partially denatured whey proteins through sonication contributed to the gel characteristics. They further postulated a three–step mechanism. Step 1 is the production of an emulsion with fine particles followed by the physical aggregation of proteins mainly through hydrophobic and Van der Waals forces at the isoelectric point (pI) of proteins upon acidification. The final step involves the formation of the 3D gel matrix with covalent cross-linking.

Almanza-Rubio et al. (2016) investigated the effects of ultrasound in modifying the textural and rheological properties of cream cheese at various power levels

(0–100 W), temperatures (4–63 °C) and time (0–30 min). The yield, spreadability and the thermostability increased due to reduced milk fat globule size and increased fat incorporation. In contrast, the increase in ultrasound power and prolonged sonication did not significantly improve the cream cheese yield. Improvements in textural and rheological properties were observed in milk thermosonicated at low power (50 W) for <30 min with temperature between 35 and 50 °C. Further, they found that protein and mineral contents did not change significantly with treatment. Strong physical forces generated by acoustic cavitation reduce the fat globule size, altered the MFGM and modified interactions with themselves and caseins. These interactions increased fat and moisture retention. The increased moisture retention weakened the firmness of the protein matrix and resulting in a softer cream cheese. Ultrasound-induced denaturation of whey proteins may also promote the formation of large aggregates with casein micelles (Sfakianakis et al. 2015) and thereby increases the water holding capacity of the cheese matrix improving the cheese yield. Bermúdez-Aguirre et al. (2010) reported that Queso Fresco cheese prepared with thermosonicated milk (129 mWml^{-1}, 63 °C for 30 min) decreased hardness and increase the springiness.

1.2.6   Crystallization

The conventional methods used for crystallization of lactose in dairy systems have long induction times due to slow crystallization rates. However, sonication is known to increase the rate of nucleation and shorten crystallization induction times (Dhumal et al. 2008). Sonocrystallization with low-frequency power ultrasound applied to aid and control crystallization is most effective when sound energy is delivered at the nucleation phase (Bund and Pandit 2007a, b). The acoustic streaming helps to initiate crystallization and control the crystallization process. Initially, crystallization takes place on existing crystals' surfaces. These crystals then act as nucleation sites. Evaporation from the internal surface of bubbles results in cooling which generates high internal supersaturation and bubbles act as nucleation sites first described by Hem (1967). Shockwaves cause further agitation and bubble disruption increasing the number of nuclei available for nucleation and the greater number of nuclei reduces crystal size, improves uniformity and increased crystallization rate. The crucial driving force for both nucleation and growth is the supersaturation level and each system, whether it is a dilute or concentrated aqueous solution shows a different response to ultrasound affecting the crystallization process. However, regardless of the medium, ultrasound prior to stirring accelerates the crystallization, decreases the size of crystals, reduces the size distribution and improves yield.

A pilot-scale study by Zisu et al. (2014) using a non-contact approach at flow rates of up to 12 L/min and energy densities varied from 3 to 16 J/mL focussed on the crystallization of lactose in whey. It was found that sonication increased the rate of crystallization and was faster compared with conventional mechanical stirring. However, once the metastable limit was reached between 120 and 180 min, the rate slowed. The crystal size distribution was narrow and the overall crystal size was small.

The study by Zisu et al. (2014) highlighted the importance of the conditions used as the yield was affected by the initial solubility of lactose. Furthermore, Dincer et al. (2014) found that ultrasound had a significant effect in reducing induction times and narrowing the metastable zone width but had no effect on individual crystal growth rate or morphology.

A considerable interest within the dairy industry with regards to crystallization of fats with the use of ultrasound prevails as controlling the crystallization of fat is a key factor governing the texture of secondary dairy products. However, the effectiveness depends entirely on the processing conditions such as the amplitude, frequency and exposure time. Martini et al. (2008) showed decreased induction times and increased generation of small crystals with the use of ultrasound for crystallization of anhydrous milk fat.

Another area that uses ultrasound for crystallization purposes is with ice creams. The size and distribution of ice crystals are quite important for proper texture and taste of the ice cream. With ultrasound, the ice crystals are small and thus impart a cream mouthfeel. In addition, the use of ultrasound distributes the ice crystals more evenly. In contrast, ultrasonic degassing can occur which may reduce the amount of entrapped air. However, Acton and Morris (1992) incorporated more gas than required to compensate for the gas content lost during ultrasound treatment. Furthermore, oxidation from radicals can impart off flavours, however, by using variable ultrasonic intensities, the detrimental issues can be overcome (Chow et al. 2003; Mortazavi and Tabatabai 2008).

1.2.7   Fat Separation

Sound waves can reflect upon themselves. These reflected sound waves can then superimpose to form 'acoustic standing waves'. High-frequency sonication (200–800 kHz) produces standing waves which have been used to rapidly cluster and flocculate oil. Phase-separated oil clusters are collected by centrifugation or skimming. Traditional and most commonly used food separation processes such as centrifugation, sedimentation, chemical induced flocculation, and membrane filtration are effective but subject to high energy consumption, excessive shear damaging product integrity, fouling limiting throughput, slow partitioning rates and requires use of additives. Low-intensity high-frequency ultrasonic separation has been used individually and in combination with traditional technologies to improve process efficiencies. Ultrasonic separation at high frequencies differs from typical low-frequency sonoprocessing as this is a much more 'gentle' processing method due to the absence of violent bubble collapses. Although this technology is not commercially used across the dairy industry, high-frequency ultrasound >400 kHz, has recently been used to recover lipids from complex mixtures at pilot scale (Juliano et al. 2013). Since this publication, high-frequency ultrasound >400 kHz, has been used to commercially to recover lipids in the Malaysian Palm Oil industry.

Standing waves position individual fat droplets at the pressure antinodes where constructive superimposition occurs or at the pressure nodes where destructive interference occurs. The antinode region is defined as the region with a high local pressure and is the point of maximum vibration. Thus, these forces act on the fat globules and other particles present in the milk system moving them towards either the node or antinode regions of the standing wave. This results in rapid agglomeration and coalescence forming large globules and phase separation. These large floccules of fat globules with increased hydrodynamic radius move rapidly to the surface according to Stokes' Law (Lamb and Caflisch 1993). This phenomenon was first observed by Miles et al. (1995) in a cuvette.

At an experimental level, ultrasonic separation as a complementary technology has been demonstrated successfully to significantly enhance the fat separation rates from milk systems by conventional methods (Leong et al. 2014a). Scale-up experiments were also demonstrated in a batch system using recombined milk at a frequency of 400 kHz (Juliano et al. 2013) and using raw whole milk at a frequency >1 MHz (Leong et al. 2014b). In addition, an 'optimal' temperature ranges (20–60°C) at which the milk fat separation rates were the greatest was trialled. Juliano et al. (2011) used frequencies of 400 kHz and 1.6 MHz generate a standing pressure wave field to destabilize fat and assist creaming in milk systems. This was previously used successfully to separate canola oil emulsions (Nii et al. 2009). Treating Recombined milk emulsions (3.5% fat) with mean diameters of 2.7 and 9.3 μm and raw milk with an emulsion size of 4.9 μm caused fat globule flocculation and clustering and increased the rate of creaming. However, creaming was most evident in raw milk as compared to recombined emulsions (Fig. 1.9).

1.3 Large-Scale Operations

Ultrasound has been utilized commercially for decades across various industries; however, it has relatively few commercial dairy applications. Much of the commercial experience has focused on sonochemistry and waste water treatment. Although applied ultrasound has widespread potential in dairy processing and despite the overwhelming interest from the research community over the past 15 years, there is little know commercial adoption of this technology by the industry. Large-scale commercial implementation in the wider food industry has been successfully demonstrated in recent years as is the example of the Malaysian palm oil industry where high-frequency ultrasound (>400 kHz) is used to recover lipids from complex coconut mixtures. In general, a number of factors are responsible for the slow uptake of this technology by the dairy industry due to novelty and limited accessibility to suitable processing equipment.

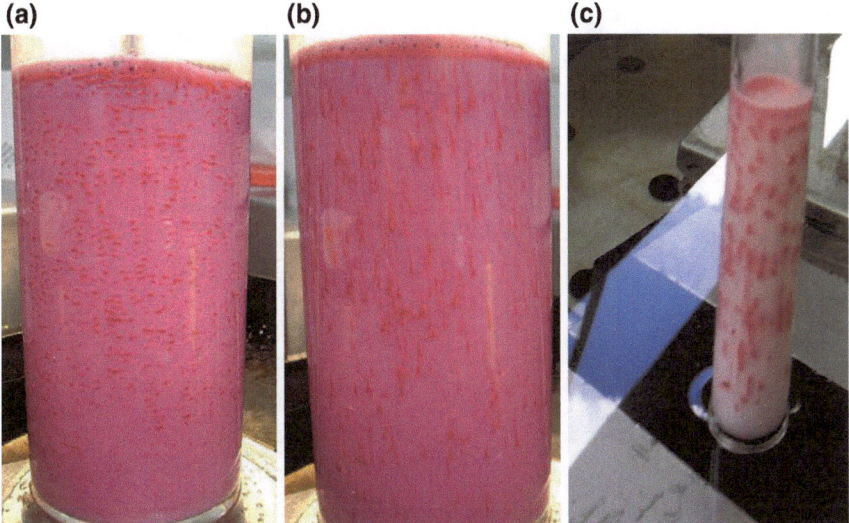

Fig. 1.9 Standing wave separation of milk fat (stained in pink) using natural whole milk at ultrasonic frequency >1 MHz (Juliano et al. 2011)

Commercial sonicators were never designed to process milk which limits the availability of commercial equipment and requires complete system design and consultation with suppliers. This adds to the complexity of adoption and increases setup costs. Traditionally ultrasonic equipment functions at low frequency (~20 kHz) and designed to function in direct contact with the treated dairy product. The common design consists of a transducer to which a sonotrode is attached (Fig. 1.10a). The titanium sonotrode emits the sound waves through air or solution and individual units can be powerful, e.g. 1–16 kW (Table 1.3; Fig. 1.10b). Commercially these may be designed in a modular arrangement and intended to operate continuously in-line with the existing facilities. A multiple sonotrode design is preferred to improve energy distribution, lower energy requirement from each sonotrode and provides an option to operated selected sonotrodes. This sonotrode design is supplied by several equipment manufacturers but pitting and degradation of the titanium will gradually occur because energy density is greatest at the sonotrode surface. A modular design allows easy and uninterrupted sonotrode replacement.

Limited options for commercial processing equipment prompted joint action in 2010 by a number of international dairy enterprises. The Dutch food research institute NIZO established an international ultrasonic processing consortium in collaboration with the dairy leaders Friesland Campina, Tetra Pak, Fonterra and others. The consortium aimed to design and develop novel in-line ultrasonic self-cleaning thermal processing equipment (NIZOvision 17). Although the outcomes of this venture are unknown, this drastic action highlights the greater need for industry-specific equipment design.

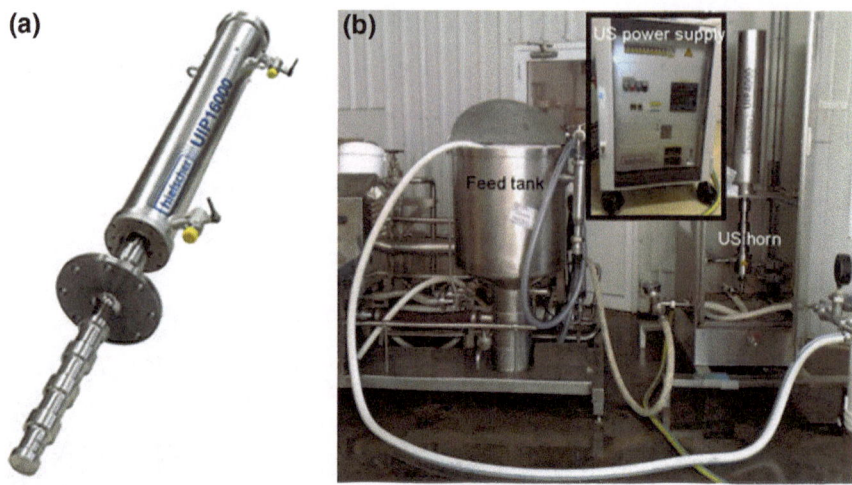

Fig. 1.10 **a** Common sonicator design consisting of titanium sonotrode and transducer; **b** 16 kW continuous operation ultrasonic flow cell (Zisu et al. 2010)

Table 1.3 Examples of low-frequency direct contact commercial sonoprocessors highlighting power output and operational flow rates

Device	Power [kW]	Freq. [kHz]	Flow-rate [m³/h]		
UIP1000hd	1.0	20	0.0	–	0.5
UIP1500hd	1.0	20	0.0	–	0.75
UIP2000	2.0	20	0.0	–	1.0
UIP4000	4.0	20	0.5	–	2.0
UIP10000	10.0	18	1.0	–	10.0
UIP16000	16.0	18			>10

Non-contact ultrasonic processing alternative exist with several equipment manufacturers now preferring to adopt this concept. This design also encourages modular implementation and in-line continuous operation. Non-contact sonication cells contain multiple low-power transducers eliminating sonotrodes, minimizing surface degradation and improving energy distribution. Ultrasound waves propagate through a metal surface while fluids are treated on the inner surface preventing direct contact. Non-contact sonication cells generating lower energy densities have been shown to initiate rapid lactose crystallization (Dincer et al. 2014). However, this setup has been successfully implemented industrially for the purpose of crystallization outside the dairy industry (Prosonix 2012).

Airborne ultrasonic processing is also a possibility and these units are generally designed with a wide flat surface and attached directly above the treatment area. Ultrasonic waves are propagated through the air and used in anti-foaming applications as documented by Chemat et al. (2011). Traditionally used for suppressing foam in the beverage industry, they also have uses in suppressing milk and whey protein foams when installed above processing tanks, storage silos, feed tanks and around filling lines.

Other ultrasonic equipment readily available includes ultrasonic spray dryers fitted with ultrasonic spray drying nozzles and various forms of ultrasonic cutters used in continuous operations for cutting and preparing cheese blocks and cheese slices, respectively (Arnold et al. 2009).

1.4 Future Trends

The radicals formed during acoustic cavitation can contribute to a variety of oxidation reactions within a dairy system especially those containing milk fat. This may exert detrimental issues within the product. However, the use of high-intensity ultrasound produces only a small number of radicals and generally considered negligible. In addition, another big concern in using high-intensity ultrasound is contamination with metal particulates from the disintegration of the ultrasonic transducer probes. A recent study by Mawson et al. (2014) revealed that metal particulates did erode from ultrasonic probes but the amount of metal particulates was below the accepted drinking water limits even after prolonged exposure. The size distribution of eroded metallic particulates was similar to those identified in existing processes (e.g. homogenization) and under no circumstances were nanoparticles detected. In addition, the use of continuous flow through non-contact industrial systems prevents contact between the ultrasonic source and the treated dairy product.

Due to the recent advances in ultrasonic dairy research, the demand for large-scale equipment compatible with industry requirements is increasing. Successful scale-up and industry adoption can only be achieved by fully understanding product and the process. The scale of the process determines the required energy outputs and these must be carefully considered when designing suitable geometries for specific products. Comprehensive understanding of the operating parameters and the expected product outcomes will ensure a successful transition to large-scale operation.

References

Abbas S, Hayat E, Karangwa M, Bashari M, Zhang X (2013) An overview of ultrasound assisted food grade nanoemulsions. Food Eng Rev 5:139–157

Abismail B, Conselier JP, Wilhelm AM, Delma H, Gourdon C (1999) Emulsification by ultrasound: droplet size distribution and stability. Ultrason Sonochem 6:75–83

Abismail B, Conselier JP, Wilhelm AM, Delma H, Gourdon C (2000) Emulsification processes: online study by multiple light scattering measurements. Ultrason Sonochem 7:187–192

Acton E, Morris GJ (1992) Methods and apparatus for the control of solidification in liquids. US Patent No. WO99/20420

Al-Hilphy ARS, Niamak AK, Al-Temimi AB (2012) Effect of ultrasonic treatment on buffalo milk homogenization and numbers of bacteria. Int J Food Sci Nutr Eng 2:113–118

Almanza-Rubio J, Gutierrez-Mendez N, Leal-Ramos M, Sepulveda D, Salmeron I (2016) Modifi-
cation of the textural and rheological properties of cream cheese using thermosonicated milk. J
Food Eng 168:223–230

Alzamora SM, Guerrero SN, Schenk M, Raffellini S, Lopez-Malo A (2011) Inactivation of microorganisms. In: Feng H, Barbosa-Canovas GD, Weiss J (eds) Ultrasound technologies for food
processing and bioprocessing. Spinger, New York

Anema SG, Klostermeyer H (1997) Heat induced, pH dependent dissociation of casein micelles on
heating reconstituted skim milk at temperatures below 100 °C. J Agric Food Chem 45:1108–1115

Arnold G, Leiteritz L, Zahn S, Rohm H (2009) Ultrasonic cutting of cheese: composition affects
cutting work reduction and energy demand. Int Dairy J 19:314–320

Arroyo C, Cebrián G, Pagán R, Condón S (2011) Inactivation of *Cronobacter sakazakii* by ultrasonic
waves under pressure in buffer and foods. Int J Food Microb 144:446–454

Ashokkumar M (2011) The characterization of acoustic cavitation bubbles—an overview. Ultrason
Sonochem 18:864–872

Ashokkumar M, Mason TJ (2007) Sonochemistry in Kirk-Othmer encyclopedia of chemical technology. Wiley

Ashokkumar M, Kentish S, Lee J, Zisu B, Palmer M, Augustin M (2009a) Processing of dairy
ingredients by ultrasonication. PCT Int Appl. WO2009/079691A1

Ashokkumar M, Lee J, Zisu B, Bhaskarcharya R, Kentish S (2009b) Sonication increases the heat
stability of whey proteins. J Dairy Sci 92:5353–5356

Barukcic I, Jakpovic K, Herceg Z, Karlovic S, Bozanic R (2015) Influence of high intensity ultrasound on microbial reduction, physic-chemical characteristics and fermentation of sweet whey.
Innov Food Sci Emerg Technol 27:94–101

Beatty N, Walsh M (2016) Influence of thermosnication on *Geobacillus stearothermophilus* inactivation in skim milk. Int Dairy J 61:10–17

Bermúdez-Aguirre D, Barbosa-Cánovas GV (2008) Study of butter fat content in milk on the inactivation of *Listeria innocua* ATCC 51742 by thermo-sonication. Innov Food Sci Emerg Technol
9:176–185

Bermúdez-Aguirre D, Mawson R, Barbosa-Cánovas GV (2008) Microstructure of fat globules in
whole milk after thermosonication treatment. J Food Sci 73:325–332

Bermúdez-Aguirre D, Mobbs T, Barbosa-Cánovas GV, Mawson R, Versteeg K (2009a) Composition properties, physicochemical characteristics and shelf life of whole milk after thermal and
thermosonication treatments. J Food Qual 32:283–302

Bermúdez-Aguirre D, Corradini MG, Mawson R, Barbosa-Cánovas GV (2009b) Modeling the
inactivation of *Listeria innocua* in raw whole milk treated under thermo-sonication. Innov Food
Sci Emerg Technol 10:172–178

Bermúdez-Aguirre D, Mobbs T, Barbosa-Cánovas GV (2010) Processing of soft Hispanic cheese
using thermosonicated milk: a study of physicochemical characteristics and storage life. J Food
Sci 75:5548–5558

Bermúdez-Aguirre D, Mobbs T, Barbosa-Cánovas GV (2011) Ultrasound applications in food
processing. In: Feng H, Barbosa-Cánovas GV, Weis J (eds) Ultrasound technologies for food and
bioprocessing. Springer, pp 65–105

Bosiljkov T, Tripalo B, Brincic M, Jezek D, Karlovic S, Jagust I (2011) Influence of high intensity
ultrasound with different probe diameter on the degree of homogenization (variance) and physical
properties of cow milk. Afr J Biotech 10:34–41

Bund RK, Pandit AB (2007a) Sonocrystalisation: effect on lactose recovery and crystal habit.
Ultrason Sonochem 14:143–152

Bund RK, Pandit AB (2007b) Rapid lactose recovery from paneer whey using sonocrystalisation:
a process optimization. Chem Eng Process 46:846–850

Caia M, Wanga S, Zheng Y, Lianga H (2009) Effects of ultrasound on ultrafiltration of *Radix
astragalus* extract and cleaning fouled membranes. Sep Purif Technol 68:351–356

Calligaris S, Plazzotta S, Bot F, Grasselli S, Malchiodi A, Anese M (2016) Nanoemulsion preparation by combining high pressure homogenization and high power ultrasound at low energy densities. Food Res Int 83:25–30

Cameron M, McMaster LD, Britz TJ (2009) Impact of ultrasound on dairy spoilage microbes and milk components. Dairy Sci Technol 89:83–98

Canselier JR, Delmas H, Wilheim AM, Abosmail B (2002) Ultrasound emulsification—an overview. J Dispersion Sci Technol 23:333–349

Carcel JA, Benedito J, Sanjuan N, Sanchez E (2009) Application of ultrasound in industry. Alimnetacion Equipos y Technol 135–141

Cavalieri F, Ashokkumar M, Grieser F, Caruso F (2008) Ultrasonic synthesis of stable and functional lysozyme microbubbles. Langmuir 24:10078–10083

Chandrapala J, Zisu B, Kentish S, Ashokkumar M (2011) Effects of ultrasound on the thermal and structural characteristics of proteins in reconstituted whey protein concentrates. Ultrason Sonochem 18:951–957

Chandrapala J, Martin GJ, Zisu B, Kentish S, Ashokkuamr M (2012a) The effect of ultrasound on casein micelle integrity. J Dairy Sci 95:6882–6890

Chandrapala J, Zisu B, Palmer M, Kentish S, Ashokkumar M (2012b) A possible mechanism to understand the ultrasound induced heat stability of whey protein concentrates. Int Non thermal Workshop, Melbourne

Chandrapala J, Zisu B, Kentish S, Ahokkumar M (2013) Influence of ultrasound on chemically induced gelation of micellar casein systems. J Dairy Res 1:1–6

Chandrapala J, Martin GJ, Kentish S, Ashokkuamr M (2014a) Dissolution and reconstitution of casein micelle containing dairy powders by high shear using ultrasonic and physical methods. Ultrason Sonochem 21:1658–1665

Chandrapala J, Zisu B, Palmer M, Kentish S, Ashokkumar M (2014b) Sonication of milk protein solutions prior to spray drying and the subsequent effects on powders during storage. J Food Eng 141:122–127

Chandrapala J, Ong L, Zisu B, Gras S, Kentish S, Ahokkumar M (2016) The effect of sonication and high pressure homogenization on the properties of pure cream. Innov Food Sci Emerg Technol 33:298–307

Chemat F, Zill-e-Huma S, Khan MK (2011) Applications of ultrasound in food technology: processing, preservation and extraction. Ultrason Sonochem 18:813–835

Cho YH, Lucey JA, Singh H (1999) Rheological properties of acid milk gels affected by the nature of the fat globule surface material and heat treatment of milk. Int Dairy J 9:537–546

Chow R, Blindt R, Chivers R, Povey M (2003) The sonocrystallisation of ice in sucrose solutions: primary and secondary nucleation. Ultrasonics 41(8):595–604

Cregenzan-Alberti O, Halpin R, Whyte P, Lyng J, Noci F (2014) Suitability of ccRSM as a tool to predict inactivation and its kinetics for *E. coli*, *S. aureus* and *P. fluorescenc* in homogeniased milk treated by manothermosonication. Food Control 39:41–48

D'amico D, Silk TM, Wu J, Guo M (2006) Inactivation of microorganisms in milk and apple cider treated with ultrasound. J Food Prot 69:556–563

Devi S, Ashokkumar M, Grieser F (2005) The influence of acoustic power on multibubble sonoluminence in aqueous solution containing organic solutes. J Phys Chem B 109:20044–20050

Dhumal R, Birandar SV, Paradkar AR, York P (2008) Ultrasound assisted engineering of lactose crystals. Pharm Res 25(12):2835–2839

Dincer TD, Zisu B, Vallet CGMR, Jayasena V, Palmer M, Weeks M (2014) Sonocrystallisation of lactose in an aqueous system. Int Dairy J 35:43–46

Earnshaw RG (1998) Ultrasound: a new opportunity for food preservation. In: Povey MJW, Mason TJ (eds) Ultrasound in food processing. Blackie Academic & Professional, London, pp 183–192

Engin B, Karagul Yuceer Y (2012) Effects of ultraviolet light and ultrasound on microbial quality and aroma-active components of milk. J Sci Food Agri 92(6):1245–1252

Ertugay MF, Sngul M, Sengul M (2004) Effect of ultrasound treatment on milk homogenization and particle size distribution of fat. Turkish J Vet Anim Sci 28:303–308

Evelyn E, Silva FVM (2015) Thermosonicatio versus thermal processing of skim milk and beef slurry: modeling the inactivation kinetics of psychrotrophic *Bacillus cereus* spores. Food Res Int 67:67–74

Freitas S, Hielscher G, Merkle HP, Gauder B (2006) Continuous contact and contamination free ultrasonic emulsification—a useful tool for pharmaceutical development and production. Ultrason Sonochem 13:76–85

Frydenberg R, Hammershoj M, Aandersen U, Greve M, Wiking L (2016a) Protein denaturation of whey protein isolates induced by high intensity ultrasound during heat gelation. Food Chem 19:415–423

Frydenberg R, Hammershoj M, Aandersen U, Greve M, Wiking L (2016b) High intensity ultrasound effects on heat induced whey proteins isolate gels depend on αLA:βLG ratio. Int Dairy J 56:1–3

Furtado G, Mantovani R, Consoli L, Hubinger M, Cunha R (2017) Structural and emulsifying properties of sodium caseinate and lactoferrin influenced by ultrasound process. Food Hydrocolloids 63:178–188

Fyfe K, Kravchuk O, Lea T, Nguyen T, Deeth H, Bhandari B (2010) Storage induced changes to high protein powders: influence on surface properties of solubility. J Sci Food Agric 91:2566–2575

Gabriel AA (2015) Inactivation of listeria monocytogens in milk by multi frequency power ultrasound. J Food Process Preserv 39:846–853

Gera N, Doores S (2011) Kinetics and mechanism of bacterial inactivation by ultrasound waves and sonoprotective effect of milk components. J Food Sci 76:M111–M119

Gondrexon N, Cheze L, Jin Y, Legay M, Tissot Q, Hengl N, Baup S, Boldo P, Pignon F, Talansier E (2015) Intensification of heat and mass transfer by ultrasound: application of heat exchanger and membrane separation processes. Ultrason Sonochem 25:40–50

Guerrero S, López-Malo A, Alzamora SM (2001) Effect of ultrasound on the survival of *Saccharomyces cerevisiae*: influence of temperature, pH and amplitude. Innov Food Sci Emerg Technol 2:31–39

Halpin R, Duffy L, Cregenzan-Alberti O, Lyng J, Noci F (2013) Combined heat treatment with mild heat, manothermosonication and pulsed electric fields reduces microbial growth in milk. Food Control 34:364–371

Heffernan S, Kelly A, Mulvihill D, Lambrich U, Schuchmann H (2011) Efficiency of a range of homogenization technologies in the emulsification and stabilization of cream liqueurs. Innov Food Sci Emerg Technol 12:628–634

Hem SL (1967) The effect of ultrasonic vibrations on crystallization processes. Ultrason 5(4):202–207

Herceg Z, Jambrak A, Lelas V, Thagard S (2012a) The effect of high intensity ultrasound treatment on the amount of *S. aureus* and *E. coli* in milk. Food Technol Biotechnol 50:46–52

Herceg Z, Juraga E, Sabota-Salamon B, Jambraka A (2012b) Inactivation of mesophillic bacteria in milk by means of high intensity ultrasound using response surface methodology. Czech J Food Sci 30:108–117

Horne DS, Davidson CM (2003) Direct observation of decrease in size of casein micelles during initial stages of renneting of skim milk. Int Dairy J 3:61–71

Hughes DE, Nyborg L (1962) Cell disruption by ultrasound. Science 138:108–114

Jambrak AR, Mason T, Lelas V, Herceg Z, Hereg L (2008) Effect of ultrasound treatment on solubility and foaming properties of whey protein dispersion. J Food Eng 86:281–287

Jambrak AR, Mason T, Lelas V, Kresic G (2010) Ultrasonic effect on physico-chemical and functional properties of α-Lactalbumin. LWT Food Sci Technol 43:254–262

Jin Y, Hengl N, Baup S, Pignon F, Gondreson N, Sztucki M, Gesan-Guiziou G, Magnin A, Abyan M, Karraouch M, Bleses D (2014) Effects of ultrasound on corss-flow ultrafiltration of skim milk: charactorisation from macro-scale to nano scale. J Membr Sci 470:205–218

Juang R, Lin K (2004) Ultrasound assisted production of w/o emulsions on liquid surfactant membrane processes. Colloids Surf, A 238:43–49

Juliano P, Kutter A, Cheng LJ, Swiergon P, Mawson R, Augustin M (2011) Enhanced creaming of milk fat globules in milk emulsions by the application of ultrasound and detection by means of optical methods. Ultrason Sonochem 18:963–973

Juliano P, Temmel S, Rout M, Swiergon P, Mawson R, Knoerzer K (2013) Creaming enhancement in a litre scale ultrasonic reactor at selected transducer configurations and frequencies. Ultrason Sonochem 20:52–62

Juliano P, Torkamani AE, Leong T, Kolb V, Watkins P, Ajlouni S, Singh TK (2014) Lipid oxidation volatiles absent in milk after selected ultrasound processing. Ultrason Sonochem 21(6):2165–2175

Juraga E, Salamon BS, Herceg Z (2011) Application of high intensity ultrasound treatment on enterobateria count in milk. Mljekarstvo 61:125–134

Kinsella J (1981) Functional properties of proteins: possible relationships between structure and function in foams. Food Chem 7(4):273–288

Knorr D, Zenker M, Heinz V, Lee D (2004) Application and potential of ultrasonics in food processing. Trend Food Sci Technol 15:261–266

Koh LLA, Chandrapala J, Zisu B, Martin GJ, Kentish S, Ashokkumar M (2014a) A comparison of the effectiveness of sonication, high shear mixing and homogenization on improving the heat stability of whey proteins solutions. Food Bioprocess Technol 7:556–566

Koh LLA, Nguyen H, Chandrapala J, Zisu B, Martin GJ, Kentish S, Ashokkumar M (2014b) The use of ultrasonic feed pre-treatment ot reduce membrane fouling in whey ultrafiltration. J Membr Sci 453:230–239

Kresic G, Lelas V, Jambrak AR, Herceg Z, Brincic SR (2008) Influence of novel food processing technologies on the rheological and thermophysical properties of whey proteins. J Food Eng 87:64–73

Lamb H, Caflisch R (1993) Hydrodynamics. Cambridge University Press

Lamminen M, Walker H, Weavers L (2004) Mechanisms and factors influencing the ultrasonic cleaning of particle fouled ceramic membranes. J Membr Sci 237:213–223

Leong T, Johansson L, Juliano P, Mawson R, McArthur S, Manasseh R (2014a) Ultrason Sonochem 21:1289

Leong T, Juliano P, Johansson L, Mawson R, McArthur SL, Manasseh R (2014b) Ultrason Sonochem 21:2092–2096

Leong TSH, Zhou M, Kukan N, Ashokkumar M, Martin G (2017) Preparation of water-in-oil-in-water emulsions by low frequency ultrasound using skim milk and sunflower oil. Food Hydrocolloids 63:685–695

Liu Z, Juliano P, Williams R, Niere J, Augustin M (2014a) Ultrasound effects on assembly of casein micelles in reconstiteud skim milk. J Dairy Res 81(2):146–155

Liu Z, Juliano P, Williams R, Niere J, Augustin M (2014b) Ultrasound improves the renneting properties of milk. Ultrason Sonochem 21(6):2131–2137

Lujan-Facundo M, Mendoza-Roca J, Cuartas-Uribe B, Alvarez-Blanco S (2016a) Cleaning efficiency enhancement of ultrasounds for membranes use din dairy industries. Ultrason Sonochem 33: 18–25

Lujan-Facundo M, Mendoza-Roca J, Cuartas-Uribe B, Alvarez-Blanco S (2016b) Study of membrane cleaning with and without ultrasound application after fouling with three model dairy solutions. Food Bioprod Process 100:36–46

Madadlou A, Mousavi ME, Emam-Djomek Z, Ehsani M, Sheehan D (2009) Sonodisruption of reassembled casein micelles at different pH values. Ultrason Sonochem 16:644–648

Marchesini G, Fasolato L, Novelli E, Balzan S, Contiero B, Montemurro F, Andrighetto I, Segato S (2015) Ultrasonic inactivation of microorganisms: a compromise between lethal capacity and sensory quality of milk. Innov Food Sci Emerg Technol 29:215–221

Martini S, Walsh MK (2012) Sensory characteristics and functionality of sonicated whey. Food Res Int 49:694–701

Martini S, Suzuki AH, Hartel RW (2008) Effect of high intensity ultrasound on crystallization behavior of anhydrous milk fat. J Am Oil Chem Soc 85:621–628

Mawson R, Rout M, Swiergon P, Ripoll Munho G, Singh T, Knoerzer K, Juliano P (2014) Production of particulates from transducer erosion: implications on food safety. Ultrason Sonochem 21(6):2122–2130

McCarthy N, Kelly P, Maher P, Fenelon M (2014) Dissolution of milk protein concentrate powders by ultrasonication. J Food Eng 126:142–148

Miles CA, Morley MJ, Hudson WR, Mackey BM (1995) Principles of separating micro-organisms from suspensions using ultrasound. J Appl Bacteriol 78:47–54

Mortazavi A, Tabatabai F (2008) Study of ice cream freezing process after treatment with ultrasound. World Appl Sci J 4(2):188–190

Muthukumaran S, Kentish S, Lalchandani S, Ashokkumar M, Mawson R, Stevens G, Grieser F (2005a) The optimization of ultrasonic cleaning procedures for dairy fouled ultrafiltration membranes. Ultrason Sonochem 12:29–35

Muthukumaran S, Kentish S, Ashokkumar M, Stevens G (2005b) Mechanisms for the ultrasonic enhancement of dairy whey ultrafiltration. J Membr Sci 258:106–114

Muthukumaran S, Kentish S, Stevens G, Ashokkumar M, Mawson G (2007) The application of ultrasound to dairy ultrafiltration: the influence of operation conditions. J Food Eng 81:364–373

Nejadmansouri M, Hosseni S, Niakosari M, Yousei G, Golmakani M (2016) Physico-chemical proeprties and storage stability of ultrasound mediated WPI stabilized fish oil nano emulsions. Food Hydrocolloids 61:801–811

Nguyen NH, Anema SG (2010) Effect of ultrasonication on the properties of skim milk used in the formation of acid gels. Innov Food Sci Emerg Technol

Nguyen NH, Anema SG (2017) Ultrasonication of reconstituted whole milk and its effect on acid gelation. Food Chem 217:593–601

Nguyen T, Lee Y, Zhou W (2012) Effect of high intensity ultrasound on carbohydratemetabolism of bifidobacteria in milk fermentation. Food Chem 130:866–874

Nii S, Kikumoto S, Tokuyama H (2009) Quantitative approach to ultrasonic emulsion separation. Ultrason Sonochem 16:145–149

Nobel S, Ross N, Protte K, Korzendorfer A, Hitzmann B, Hinrichs J (2016) Microgel particle formation in yoghurt as influenced by sonication during fermentation. J Food Eng 180:29–38

Noci F, Walking-Ribeiro M, Cronin D, Morgan DJ, Lyng JG (2009) Effect of thermosonication, pulsed electric field and their combination on inactivation of *L. innocua* in milk. Int Dairy J 19:30–35

O'Sullivan J, Arellano M, Pichot R, Norton I (2014) The effect of ultrasound treatment on the structural, physical and emulsifying properties of dairy proteins. Food Hydrocolloids 42:386–396

O'Sullivan J, Murray B, Flynn C, Norton I (2015a) Comparison of batch and continuous ultrasonic emulsification processes. J Food Eng 167:114–121

O'Sullivan J, Beevers J, Park M, Greenwood R, Norton I (2015b) Comparative assesement of the effect of ultrasound treatment on protein functionality pre- and post emulsification. Colloids Surf, A 484:89–98

Ordóñoz JA, Aguilera MP, Garcia ML, Sanz B (1987) Effect of combined ultrasonic and heat treatment on the survival of a strain of *Staphylococcus aureus*. J Dairy Res 54:61–67

Pagán R, Mañas P, Alvarez I, Condón S (1999) Resistance of *Listeria monocytogenesto* ultrasonic waves under pressure at sublethal (manosonication) and lethal (manothermosonication) temperatures. Food Microb 16(2):139–148

Prosonix (2012) Revolutionizing respiratory medicine (www.prosonix.co.uk). Website visited 20 December 2012.

Raso J, Palop A, Condon S (1998) Inactivation of *Bacillus subtilis* spores by combining ultrasonic waves under pressure and mild heat treatment. J Appl Micro 85:849–854

Riener J, Noci F, CroninDA Morgan DJ, Lyng G (2009) The effect of thermosonication of milk on selected physicochemical and microstructural properties of yoghurt gels during fermentation. Food Chem 114:905–911

Riener J, Noci F, CroninDA Morgan DJ, Lyng G (2010) A comparison of selected quality characteristics of yoghurts prepared from thermosoicated and conventially heated milks. Food Chem 119:1108–1110

Ross AIV, Griffiths MW, Mittal GS, Deeth HS (2003) Combining nonthermal technologies to control foodborne microorganisms. International J Food Microb 89(2–3):125–138

Saffon M, Britten M, Pouliot Y (2011) Thermal aggregation of whey proteins in the presence of butter milk concentrate. J Food Eng 103:244–250

Sengul M, Erkaya T, Balsar M, Ertugay F (2011) Effect of photosonication treatment on inactivation of total and coliform bacteria in milk. Food Control 22:1803–1806

Sfakianakis P, Tzia C (2010) Yoghurt from ultrasound treated milk: monitoring of fermentation process and evaluation of product quality characteristics. Food Process Eng ICEF11 Proc 3:1649–1654

Sfakianakis P, Topakas E, Tzia C (2015) Comparative study on high intensity ultrasound and pressure milk homogenization: effect on the kinetics of yoghurt fermentation process. Food Bioprocess Technol 8:548–557

Shamila-Syuhada A, Chuah L, Wan-Nadiah W, Cheng L, Alkarthi A, Effarizah M, Rusul G (2016) Inactivation of microbiota and selected spoilage and pathogenic bacteria in milk by combinations of ultrasound, hydrogen peroxide and active lactoseperoxidase system. Int Dairy J 61:120–125

Shanmugam A, Ashokkumar M (2014a) Ultrasonic preparation of stable flax seed oil emulsions in dairy systems—physicochemical characterization. Food Hydrocolloids 39:151–162

Shanmugam A, Ashokkumar M (2014b) Functional properties of ultrasonically generated flaxseed oil-dairy emulsions. Ultrson Sonochem 21:1649–1657

Shanmugam A, Chandrapala J, Ashokkumar M (2012) The effect of ultrasound on the physical and functional properties of skim milk. Innov Food Sci Emerg Technol

Shen X, Fang T, Gao F, Guo M (2017) Effects of ultrasound treatment on physicochemical and emulsifying properties of whey proteins pre and post thermal aggregation. Food Hydrocolloids 63:668–676

Sivakumar M, Senthilkumar P, Majumdar S, Pandit AB (2002) Ultrasound mediated alkaline hydrolysis of methyl benzoate reinvestigation with crucial parameters. Ultrason Sonochem 9:25–30

Stathopulos PB, Scholz GA, Hwang YM, Rumfeldt JA, Lepock JR, Meiering EM (2004) Sonication of proteins causes formation of aggregates that resemble amyloid. Protein Sci 13:3017–3027

Tan M, Chin N, Yusof Y, Taip F, Abdullah J (2015) Characterisation of improved foam aeration and rheological properties of ultrasonically treated whey protein suspension. Int Dairy J 43:7–14

Vankova N, Tcholakova S, Denkov ND, Ivanov IB, Vulchev VD, Danner T (2007) Emulsification in turbulent flow—1. Mean and maximum drop diameter in inertial and viscous regimes. J Colloid Int Sci 312:363–380

Venugopal V (2008) Marine products for healthcare: functional and bioactive nutraceutical compounds from the ocean. CRC Press, Boca Raton, pp 51–102

Vercet A, Oria P, Quina P, Crelier S, Lopez P (2002) Rheological properties of yoghurt made with milk submitted and manothermosonication. J Agric Food Chem 50:6165–6171

Villamiel M, de Jong P (2000) Influence of high intensity ultrasound and heat treatment in continuous flow on fat, protein and native enzymes of milk. J Agric Food Chem 48:472–478

Wang D, Sakakibara M (1997) Lactose hydrolysis and B-galactosidase activity in sonicated fermentation with *Lactobacillus strains*. Ultrason Sonochem 4:255–261

Weiss J, Kristbergsson K, Kjartansson GT (2011) Engineering food ingredients with high intensity ultrasound. In: Feng H, Barbosa-Cánovas G, Weiss J (eds) US technologies for food and bio processing. Springer, New York, pp 239–285

Wu VC (2008) A review of microbial injury and recovery methods in food. Food Microb 25:735–744

Wu H, Hulbert J, Mont JR (2000) Effect of ultrasound on milk homogenization and fermentation with yoghurt starter. Innov Food Sci Emerg Technol 1:211–218

Wu H, Hulbert GJ, Mount JR (2001) Effects of ultrasound on milk homogenization and fermentation with yoghurt starter. Innov Food Sci Emerg Technol 1:211–218

Yanjun S, Jianhang C, Shiwen Z, Hongjuan L, Jing L, Lu L, Uluko H, Yanling S, Wenming C, Wupeng G, Jiaping L (2014) Effect of power ultrasound pretreatment on the physical and functional properties of reconstituted milk protein concentrate. J Food Eng 124:11–18

Ye A, Anema SG, Singh H (2004) High pressure induced interactions between milk fat globule membrane proteins and skim milk proteins in whole milk. J Dairy Sci 87:4013–4022

Zhao L, Zhang S, Uluko H, Lu L, Xue H, Kong F, Lv J (2014). Effect of ultrasound pre-treatment on rennet-induced coagulation properties of goat's milk. Food Chem 165:167–174

Zisu B, Bhaskarcharya R, Ashokkumar M, Kentish S (2010) Ultrasonics processing of dairy systems in large scale reactors. Ultrason Sonochem 17:1075–1087

Zisu B, Lee J, Chandrapala J, Bhaskarcharya R, Palmer M, Kentish S, Ashokkumar M (2011) Effect of ultrasound on the physical and functional properties of reconstituted whey protein powders. J Dairy Res 78:226–232

Zisu B, Schleyer, Chandrapala J (2012) Application of ultrasound to reduce viscosity and control the rate of age thickening of concentrated skim milk. Int Dairy J 1–3

Zisu B, Sciberras M, Jayasena V, Weeks M, Palmer M, Dincer T (2014) Sonocrystallsiation of lactose in concentrated whey. Ultrason Sonochem 21(6):2117–2121